METAMORPHS

Transforming Mathematical Surprises

Robert Byrnes

METAMORPHS 1 - 5: **Symmetrical diamorphs**

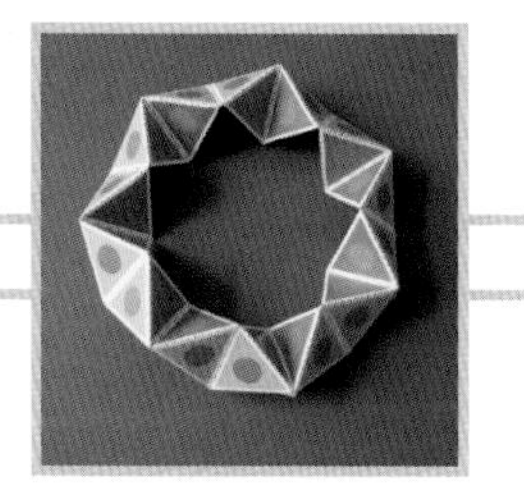

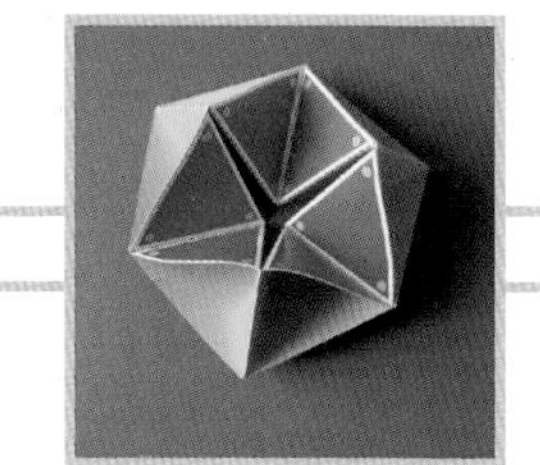
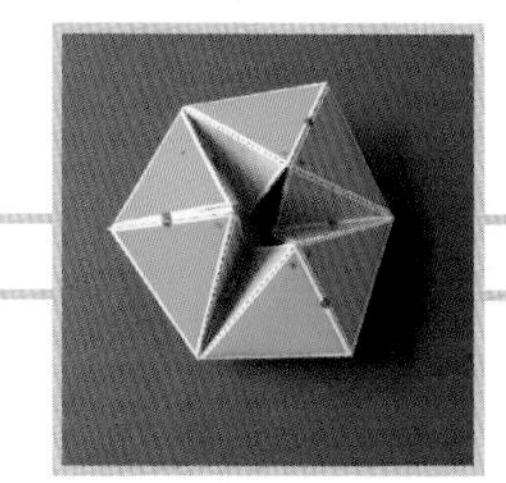
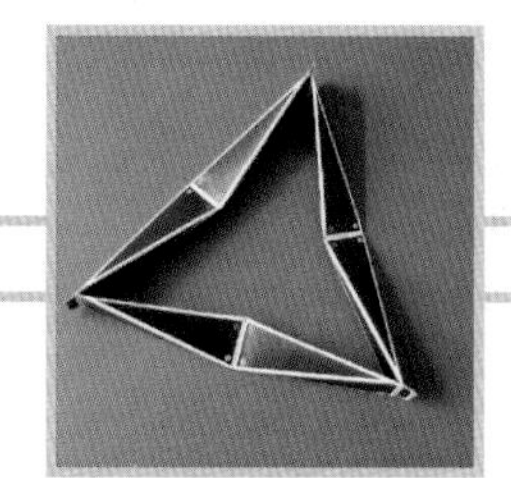

METAMORPHS 6 & 7: **Asymmetrical diamorphs**

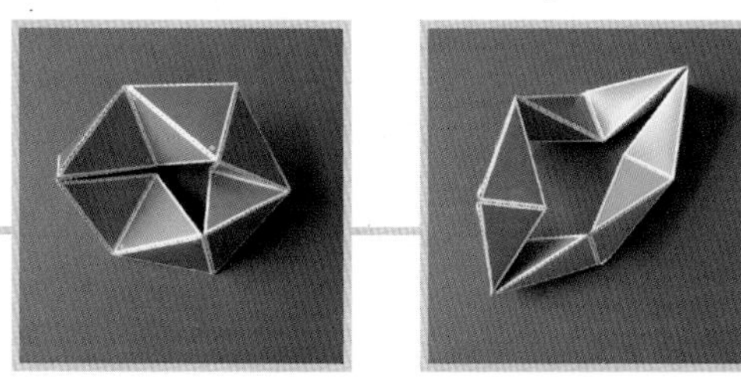

METAMORPHS 8 - 11: **Plane metamorphs**

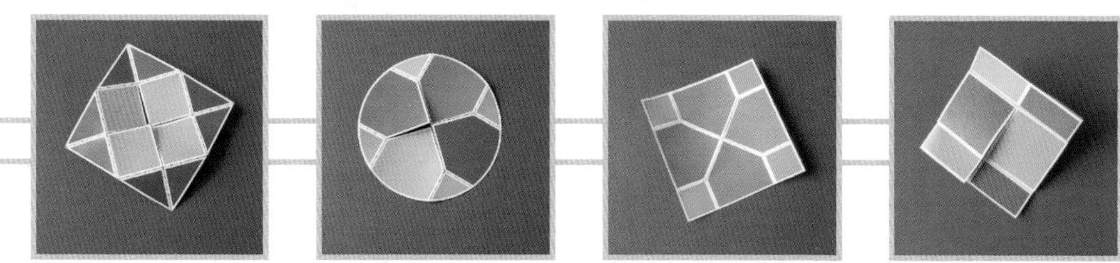

METAMORPHS 12 - 14: **Orthomorphs**

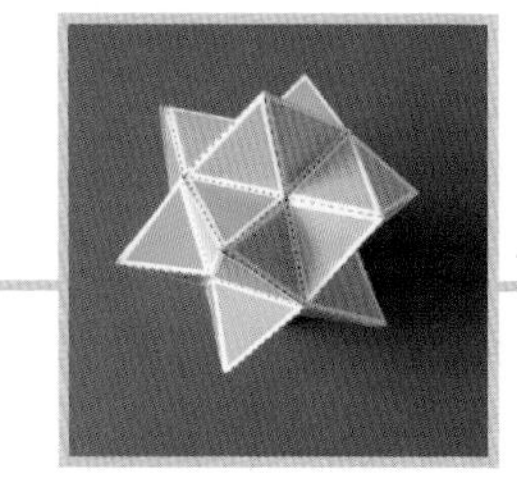

Tarquin Publications

METAMORPHS

The word 'metamorph' is an invented one to describe these fascinating curiosities, but since it comes directly from the word metamorphosis, it is clear that they are physical objects which change their form as they are manipulated. The models in this collection have been chosen from the infinity of possible metamorphs both to illustrate the principles that they obey and also to show how beautiful, elegant and interesting such models can be.

There are two main types of metamorph, diamorphs and orthomorphs. The words describe how they progress.

Diamorphs pass through their own centre in some way. This also is an invented word but it comes from the Greek 'dia', meaning through. All their hinges are activated at the same time and they flow through their centres in such a smooth and harmonious way that people who have not seen such models before find hard to believe.

Orthomorphs also pass through their centres but this is not usually immediately obvious or apparent. A much more obvious feature of orthomorphs is that their motion is step-like and the completion of each step serves to bring new hinges into a position where they can be activated and allow the motion to continue. Because they open along alternate axes at right angles, the invented word orthomorph has been derived from the Greek 'orthos', meaning straight or right.

In this collection, there are five examples of symmetrical diamorphs, two examples of asymmetrical diamorphs, three orthomorphs and four flat two-dimensional metamorphs, one of which is a diamorph and the other three are orthomorphs.

Cut out and make the models and read the text about these interesting moveable forms. It is recommended that you make the models in the order given.

Assembly instructions

Scoring
Scoring is very important if you are to make accurate models. It makes the paper fold cleanly and accurately along the line you want. Use a ball point pen which has run out of ink and rule firmly along the fold lines. Experienced model makers may use a craft knife, but it needs care not to cut right through the paper.

1. Cut out the model of your choice from the book, keeping well away from the outline.
2. Score along all fold lines and then cut out the individual pieces precisely.
3. Carefully fold along the fold lines to make either a hill fold, a valley fold or a hinge fold.

Solid cut line — Dashed and dotted fold lines

Fold away from you to make a hill fold — Fold towards you to make a valley fold — Fold both ways to make a hinge fold

4. Crease firmly.
   Do not start to glue until all the creases are done.

Type of Glue
To get the best results you need a glue which sets quickly but not instantly and which does not leave dirty marks. We particularly recommend a petroleum based glue such as 'UHU All Purpose', especially the 'gel' version. Both 'BOSTIK Clear' and white latex adhesives such as 'COPYDEX' also give good results.

5. There are letters and numbers on the flaps which show the best order in which to glue the models together. If the model is made up from several units then certain flaps are printed with letters of the alphabet. Work first in alphabetical order A, B, C ...
6. Next, glue the flaps marked with numbers to make the model or link the units together. Work in numerical order 1, 2, 3 ... Glue flaps with the same letter or number at the same time.
7. Match arrows and triangles to orientate the pieces correctly.
8. If certain flaps seem difficult to glue, try rotating the model gently into another position. This will make it easier to apply pressure in the right direction.
9. Finally allow the glue to dry completely. Once it is dry, start by gently turning the model slowly and carefully through its sequence. After a few turns the model will become 'run in', turn with ease and last for a very long time.

© 2004: Robert Byrnes
I.S.B.N: 1 899618 60 0
Design: Magdalen Bear
Printing: Fuller-Davies, Ipswich

CE

All rights reserved

Tarquin Publications
Stradbroke
Diss
Norfolk IP21 5JP
England

What makes a model a metamorph?

To be a metamorph, a model has to step or flow through a sequence of different shapes before returning to its original starting position. There is the important proviso that the movement, the sequence of steps of the transformation, always continues in the same direction. There are never any 'blind alleys' in the folding or turning process where the only way to carry on is to reverse back out the way you came in.

The models in this collection have been carefully chosen to illustrate the principles and possibilities inherent in metamorphs. It is recommended that you cut them out and make them in the order given. Then read the text with the appropriate models to hand. This is certainly a topic where the theory is much easier to understand with a visual and tactile aid close by. That being said, there are mathematical asides and calculations which will not necessarily be of interest to everyone. Those can be safely skipped or delayed until you want to design metamorphs for yourself.

Metamorph 1

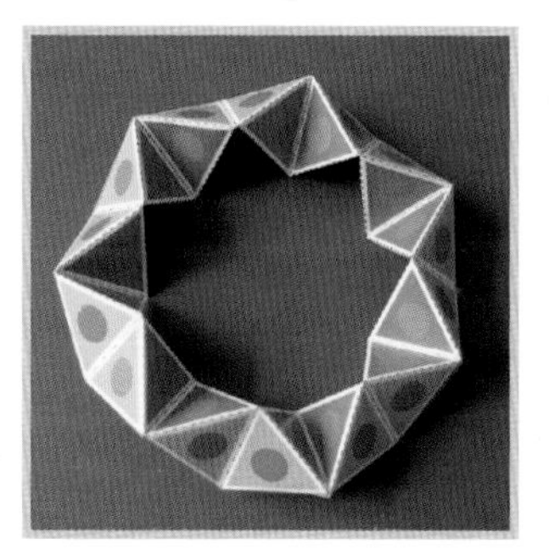

This is a good model to start with. All its faces are equilateral triangles and each of the fourteen elements in the rotating ring is a regular tetrahedron. Observe how the tetrahedra are hinged together in pairs and that alternate hinges are at right angles to each other.

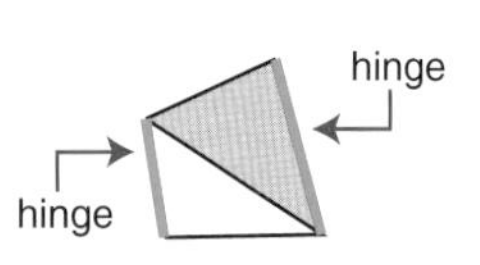

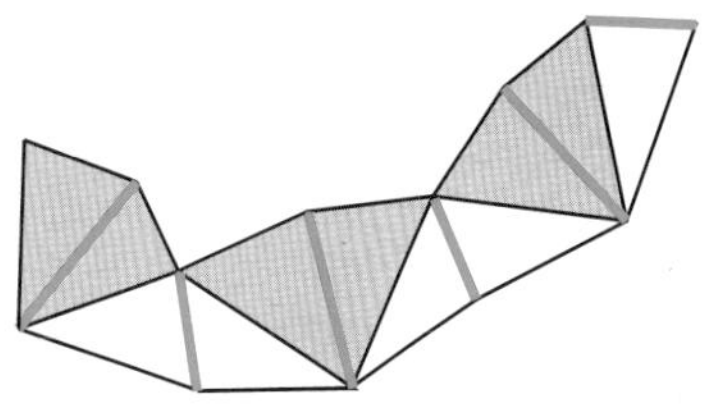

It is most important to look at the model and be sure that you appreciate what is meant when it says that the hinges are at right angles or perpendicular. Because the lines are not in the same plane, they do not intersect and so it is not always immediately obvious whether or not they are at right angles.

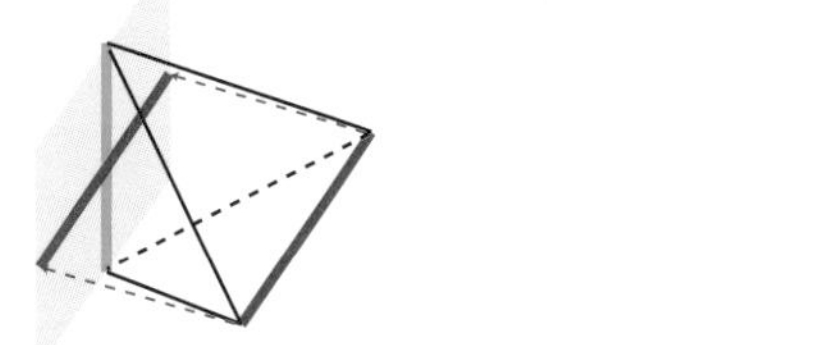

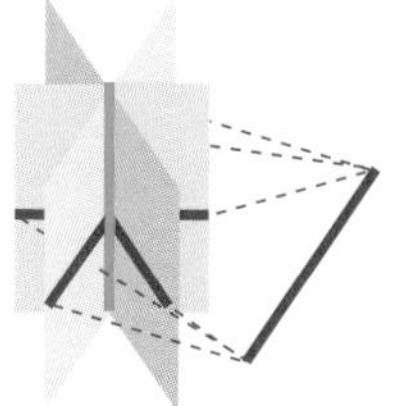

Mathematicians deal with this problem by projecting one of the lines on to a plane passing through the other. If this projected line is perpendicular to the original line, then the two lines in space are perpendicular. Note also that through any line there is an infinity of planes to project on to and that the projected line and the original will be perpendicular in every one of those planes.

Now check that the model can be rotated through its centre indefinitely. It is the flexing of the hinges backwards and forwards which allows this to happen. Such metamorphs with n regular tetrahedra are also known as 'rotating rings' for rather obvious reasons. and it has been proved elsewhere that if $n \geq 6$, then a ring is possible and that if $n \geq 8$, it will rotate. If $n \geq 22$ then the ring can be knotted and will still rotate. It is still a metamorph.

Another important feature to note is that the model is rather floppy and can be easily deformed while it is being rotated. While this does not affect its theoretical properties as a metamorph, in general people prefer metamorphs which are more constrained in their motion. They seem altogether more satisfying. Metamorph 2, the next model to make, is one where $n = 8$. It is the most constrained of the rotating rings which can be made from regular tetrahedra.

Metamorph 2

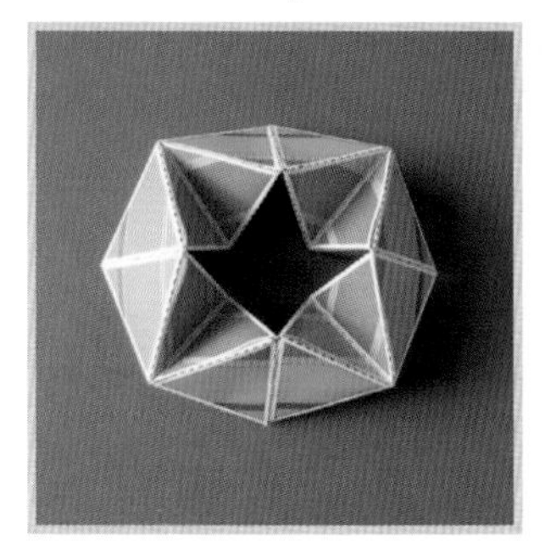

This model with eight elements is both the simplest of the regular tetrahedra rotating rings and the least deformable. It can be deformed a little but easily springs back into a pleasingly symmetrical shape. Note the four-pointed stars which form and reform as it is turned and how the model becomes more symmetrical in what can be called the 'resting' positions.

If the model is resting on a level horizontal surface there are positions when four of the hinges are vertical and four are horizontal.

When n = 6, the ring will not rotate

This model is so constrained that it is not a metamorph and will not turn through its centre. The best that it can be made to do is to rock back and forth between two limiting positions. Some people may be able to visualise this situation in three dimensions but for most of us a physical model is a great help to understanding.

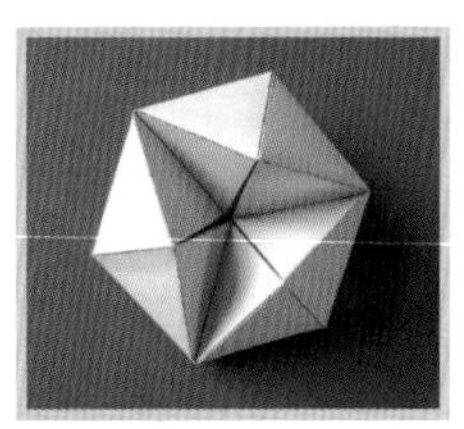

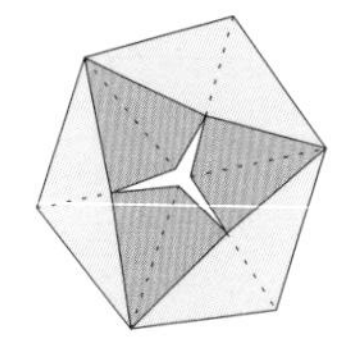

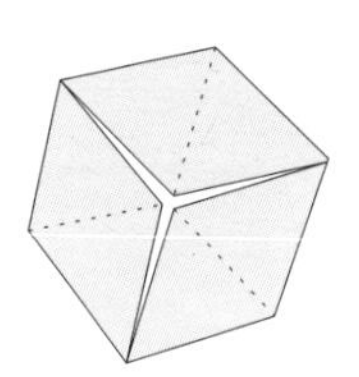

It is recommended that you draw the net which is given here. All the angles are either 30°, 60° or 90°.

It is best to begin by glueing the flaps marked 1, 2, 3 and then inserting the end flaps to complete the six-element ring. Then see for yourself what happens.

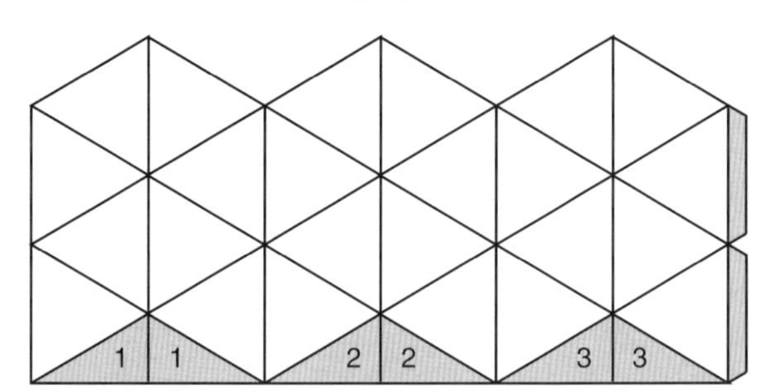

It is a great help in our appreciation of the properties of metamorphs to understand why this particular model cannot be a metamorph. All the angles of the regular tetrahedra are 60°, and when the six vertices come together at the centre they make a total of 360°. One might therefore expect the surface to become flat and for the vertices to just squeeze through. However the model shows that they do not. Is this just a consequence of the thickness of the paper? The answer is an emphatic no!

There is something much more fundamental at work here and the explanation is both simple to visualise and revealing.

The photograph on the right shows six regular tetrahedra lying on a flat horizontal surface. Their six 60° angles do meet and no gaps are left, so once again we have a total angle of 360°. However, it can be seen that no other pairs of edges meet and so there can be no other hinges. The elements cannot have hinges at opposite ends as is required for metamorphs if they are to attain this position.

It is as simple as that. A rotating ring with only six regular tetrahedra cannot rotate, even theoretically. Only such models where $n \geq 8$ will rotate and be true metamorphs.

Full cycle mobility

Another way to describe the motion of a metamorph is to say that it has 'full-cycle mobility' It is evident that a model with six regular tetrahedral elements does not have this essential property whereas both models 1 and 2 do. What we need to discover are models where the motion is sufficiently constrained to be interesting, yet not so constrained as to fail to be metamorphs.

We are therefore looking for interesting constrained models and will start with those that have six elements. Of those, we will study six-element metamorphs where the triangular faces are congruent. For Metamorph 3, the faces are chosen to be isosceles triangles.

Metamorph 3

Models of this type where all the faces are isosceles triangles are known as 'kaleidocycles' after the Greek word 'kalos' for beauty. The proportions of the sides of the triangles can be so chosen so that the vertices just come together to 'fill the centre' as the model is rotated. When the elements have this property, as they do in this model, they just touch each other before squeezing through. They do this four times in each complete cycle.

An additional feature of this particular model is that six of the faces have been removed to allow the interior to be seen at various stages during the rotation. It takes on a flower-like form that can be enhanced by the choice of colours for their decoration.

Kaleidocycles have been known for some time and the collection in the larger photograph above shows the beauty that such models can offer. During their rotations they all have outlines of regular hexagons for the six-tetrahedron ones and squares for eight-tetrahedron ones. Those regular geometrical shapes appear in different orientations.

On the right are two 'polyhedra flowers' which can best be described as 'even more cut-away' kaleidocycles. They are also true metamorphs as they appear to bloom, grow and then turn inside-out as the model is manipulated.

Metamorph 4

This model goes one step further than a kaleidocycle towards a more general kind of model metamorph because all its faces are scalene triangles, not isosceles ones. However, it is true that all the triangles are still congruent and that they always appear on the model as pairs of mirror images. Note how the hinges are lines of mirror symmetry between such pairs. A particularly interesting feature of this model is that six vertices meet and touch at a single point four times during a cycle. When they do, they alternately define a plane surface on each side. During a complete cycle all four colours complete such a plane, blue and green on one side and yellow and lime on the other.

For this model and also model 3, the calculations to make sure that they both had the required properties were done by the author. The next step is to explain how to do such calculations and how to know in advance if a model will have full-cycle mobility.

A single element, the elemental length

The photograph, diagram and net below show a single element taken from model 4, with the hinges and their mid-points clearly marked. The line joining those mid-points passing through the space inside the element is called the 'elemental length'. It is also the 'common normal' because it meets both hinges at right angles. This elemental length is the shortest distance between the two hinges and is a fundamental distance in calculations.

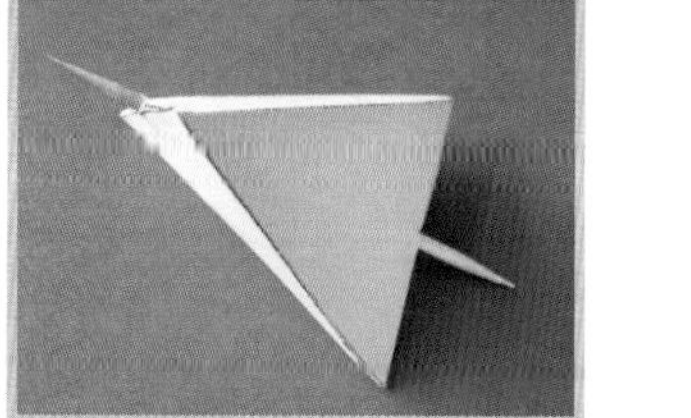

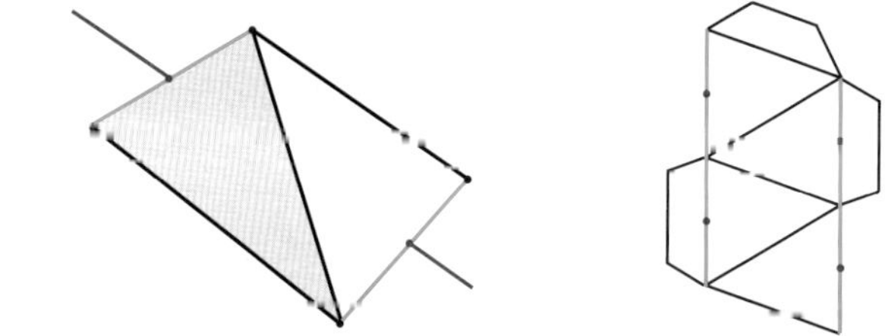

It is a good idea to make such a model for yourself and to push a cocktail stick through it to represent the elemental length. You will clearly see that the stick makes a right angle as it emerges from the centre of each hinge.

The common normal is always the shortest distance between two lines in space but in general it will not join the mid-points of the lines. The fact that it does in the case of all the metamorphs in this collection is because the decision was taken to restrict ourselves to certain kinds of models; those where all the faces are congruent and where the nets of the individual elements are similar to the one above. It is exactly this type of restriction and arrangement which has been found to produce the most beautiful and most symmetrical kinds of metamorphs.

If we decide to abandon these restrictions and create irregular metamorphs, then there is much original research to be done and discoveries to be made. However, such models lie outside the scope of this particular book.

The elemental length of a regular tetrahedron

Let us now calculate the elemental length e for a regular tetrahedron of side 2h. It is easier to use h for the semi-hinge length and so avoid lots of fractions.

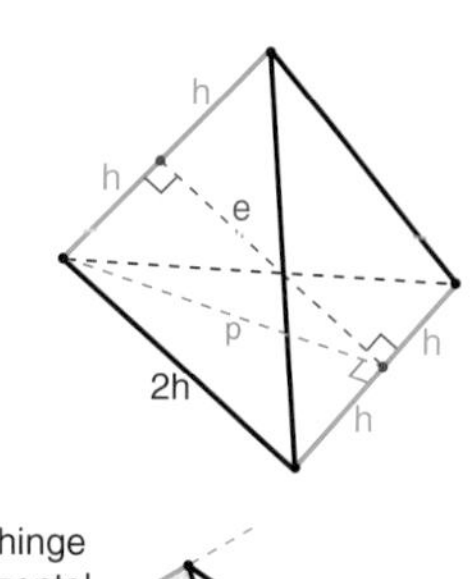

Using the right angled triangles in the diagram, which shows the inside of a tetrahedron.

$p^2 = 4h^2 - h^2 = 3h^2$ and $e^2 = p^2 - h^2$ giving $e^2 = 3h^2 - h^2 = 2h^2$

Hence $e = h\sqrt{2}$

The elemental length is $\sqrt{2}$ times the semi-hinge length.

The angle of 'twist', ϕ

Another important parameter of these tetrahedra is the angle that a hinge twists relative to the hinge at the other end of the element. It can be seen if you hold the element so that you are looking along the elemental length with the far hinge held horizontally. The near hinge slopes relative to the horizontal hinge and the angle it makes it called ϕ (phi).

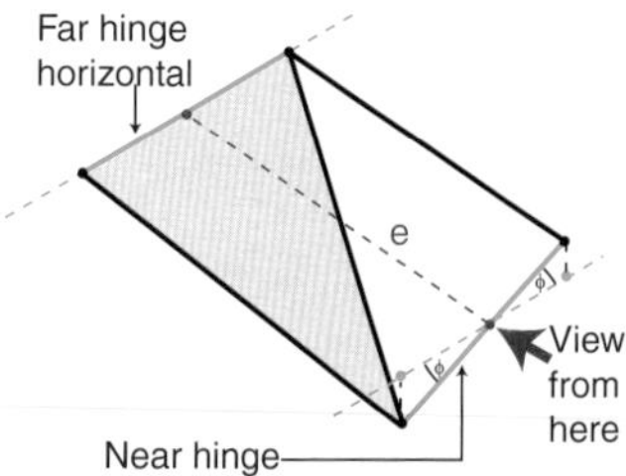

It is very much worthwhile looking for this angle on model 4. Hold it so that one of the hinges is horizontal. Then turn the nearest element so that you are looking along the direction of its elemental length. The nearer hinge slopes relative to the horizontal hinge at the angle ϕ. For this model $\phi = 75°$ approximately.

Although ϕ is conventionally measured positive in an anticlockwise direction, it is shown in the following diagrams as clockwise in order to simplify them. For a regular tetrahedron the value of $\phi = 90°$. You can confirm this by looking at models 1 and 2.

Calculating the sides and angles of an element

Once the lengths of the semi-hinge h, the elemental length e and the angle of twist ϕ are known, all the sides and angles of an element can be calculated. All the triangular faces are congruent and two of them are mirror images of the other two.

Using a standard notation, one side of the triangle is 2h and the others are a and b. The angles opposite those sides are A and B with C being the angle opposite the hinge.

To obtain algebraic relationships between the lengths we use three-dimensional coordinates. One hinge is taken as the y axis with its mid-point the origin. The elemental length is then rotated to become the x axis. The z axis is perpendicular to the xy plane.

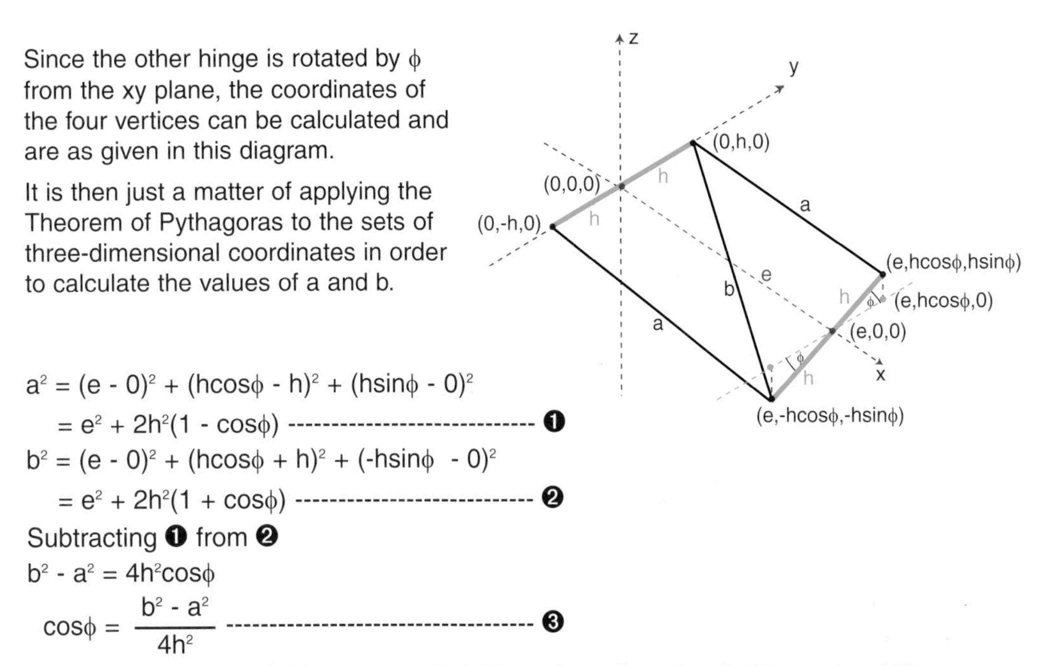

Since the other hinge is rotated by ϕ from the xy plane, the coordinates of the four vertices can be calculated and are as given in this diagram.

It is then just a matter of applying the Theorem of Pythagoras to the sets of three-dimensional coordinates in order to calculate the values of a and b.

$$a^2 = (e - 0)^2 + (h\cos\phi - h)^2 + (h\sin\phi - 0)^2$$

$$= e^2 + 2h^2(1 - \cos\phi) \text{ ---------------------------- } ❶$$

$$b^2 = (e - 0)^2 + (h\cos\phi + h)^2 + (-h\sin\phi - 0)^2$$

$$= e^2 + 2h^2(1 + \cos\phi) \text{ --------------------------- } ❷$$

Subtracting ❶ from ❷

$$b^2 - a^2 = 4h^2\cos\phi$$

$$\cos\phi = \frac{b^2 - a^2}{4h^2} \text{ ---------------------------------- } ❸$$

When b = a, the model becomes a kaleidocycle and $\cos\phi = 0$. Hence $\phi = 90°$. You can confirm this point by looking at Metamorph 2.

Adding ❶ and ❷

$$a^2 + b^2 = 2e^2 + 4h^2$$

$$e^2 = \frac{a^2 + b^2 - 4h^2}{2} \text{ ---------------------------- } ❹$$

When a = b = 2h, the element is a regular tetrahedron and $e^2 = 2h^2$, confirming the earlier result that $e = h\sqrt{2}$

The angles A, B, C can be calculated using the Cosine and Sine Rules from ❹ $e^2 = \frac{2ab\cos C}{2}$. Hence $\cos C = \frac{e^2}{ab}$. Also $\frac{\sin A}{a} = \frac{\sin B}{b} = \frac{\sin C}{2h}$

Working with unit lengths and proportions

Since the actual size of the model simply is a matter of convenience, it is more useful to define the elemental length as one unit and to express the other lengths as proportions. It is then a question of multiplying these proportions by a convenient constant to produce a model of the size required.

For model 4, the given values are $\cos\phi = 0.25$, $\phi = 75.5225°$, e = 1,

giving: a = 1.18322, b = 1.29099, h = 0.5164, A = 60°, B = 70.8934°, C = 49.1066°.

Linking the elements together to make a ring

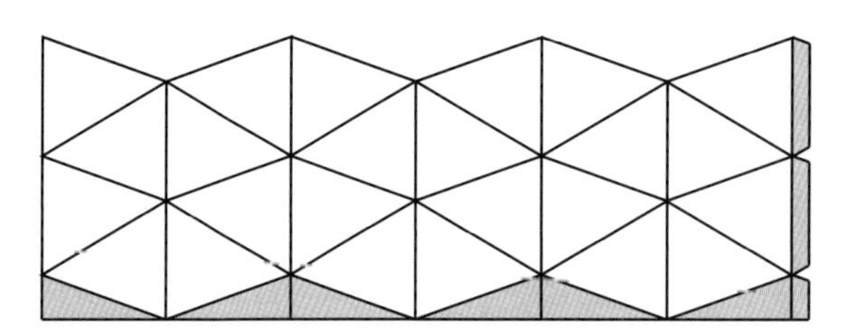

We are concentrating on investigating six-element metamorphs and the drawings above show a generalised net for such models. The photographs below show three typical models made using just such a net.

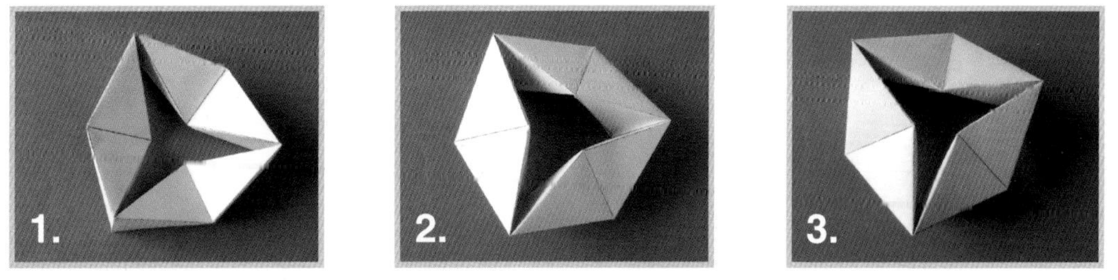

These rather similar looking models have all been made using measurements in different proportions but only 1 and 3 are metamorphs and have full cycle mobility. Model 2 does not and is an example of a rotating ring that does not rotate. Our next investigation is to find out why this might be and how to know in advance into which class a model will fit.

The idea of 'maximum hinge length'

Common sense will tell us that if a rotating ring will not rotate, it is because in some sense there is simply too much paper in the way for the tetrahedra to squeeze through the centre. Both models 3 & 4 just squeeze through and that undoubtedly is an important part of their attraction. There is always a satisfaction in going right to the limit.

At the most critical part of the rotation, the hinges come together and if they are too long, then they simply will not pass through. If they are shorter than some limiting value, there will still be a gap in the middle at the tightest point. This limiting value 2H is known as the 'maximum hinge length'. Amazingly, for this kind of metamorph the value of H, the maximum semi-hinge length depends only on ϕ, the angle of twist. The full proof is given in appendix A on page 16. Here is the result in all its glorious simplicity.

$$H = \sqrt{\cot^2 60^\circ - \cot^2\phi} \quad \text{where it is evident that } 60^\circ \leqslant \phi \leqslant 90^\circ$$

This form applies only to the rather symmetrical six-element rotating ring of the sort we are discussing. For a general 2n rotating ring of this type, the expression becomes:

$$H = \sqrt{\cot^2 \frac{180^\circ}{n} - \cot^2\phi} \quad \text{where } \frac{180^\circ}{n} \leqslant \phi \leqslant 90^\circ$$

For an eight-element metamorph, n = 4 and ϕ lies between 45° and 90°.

The parameters for the three models above are given below. Since ϕ = 70°, we are able to

Model 1: ϕ =70° e = 1 h = 0.39
A = 63.83° a = 1.0955 B =76.45° b = 1.1867 C = 39.72°.

Model 2: ϕ =70° e = 1 h = 0.51
A = 58.32° a = 1.1150 B = 73.16° b = 1.3031 C = 48.52°

Model 3: ϕ =70° e = 1 h = 0.43
A = 61.85° a = 1.1150 B =75.30° b = 1.2232 C = 42.85°

Setting up a spreadsheet

It can be rather tedious calculating the maximum hinge length, the sides and the angles of such tetrahedra even with the aid of a calculator. It is also easy to make mistakes. However, this is a perfect situation in which to set up a spreadsheet or write a program to automatically calculate a, b, A, B, C and H for any values of ϕ and h.

Once such a spreadsheet is constructed or program written, it can be tested on the known models given here. It can then be used to construct the nets of metamorphs of your own, knowing in advance that they will have the properties that you require.

Calculating maximum hinge lengths

As we have seen, the expression for H implies that ϕ can only lie between 60° and 90°. The table below, calculated using a spreadsheet, shows the stages in the calculation at intervals of 5° between these limits. Remember that e = 1.

ϕ	$\cot\phi$	$\cot^2 60 - \cot^2\phi$	H
60	0.5774	0.0000	0.0000
65	0.4663	0.1159	0.3404
70	0.3640	0.2009	0.4482
75	0.2679	0.2615	0.5114
80	0.1763	0.3022	0.5498
85	0.0875	0.3257	0.5707
90	0.0000	0.3333	0.5774

As there is only one variable ϕ, the values of H can be shown in a single graph.

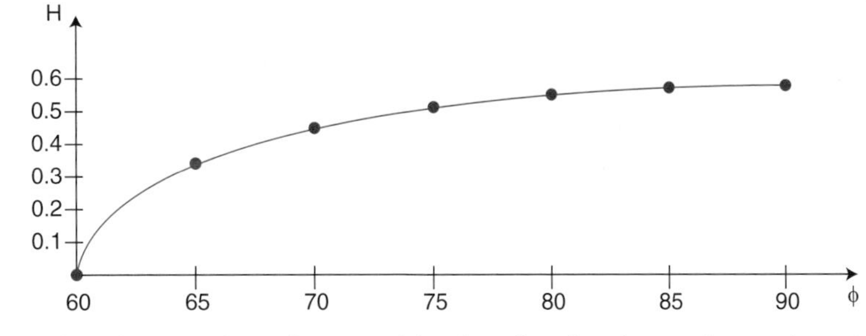

We have already seen that when ϕ = 90°, the triangles that make up the tetrahedra are either equilateral or isosceles. At the other limit, when ϕ = 60°, it can be seen that the maximum hinge length is zero and the elements have to be infinitely narrow and thus cannot be made out of paper. Metamorph 5 explores what happens to the tetrahedra when it is close to that lower limit, namely when ϕ = 61°.

Metamorph 5

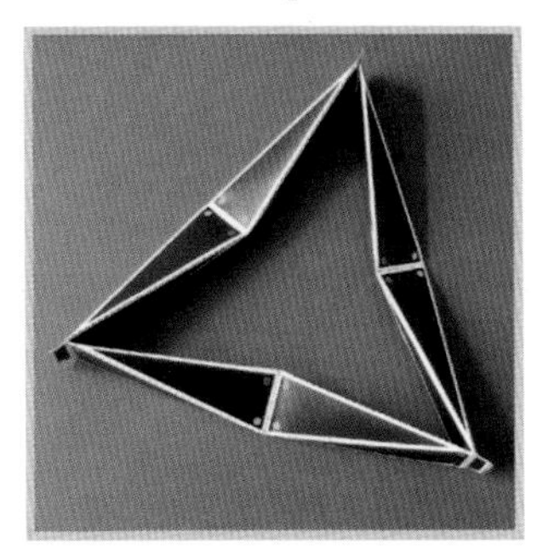

Using the spreadsheet or a calculator to work out the maximum hinge length H for the values of ϕ =61° and e = 1, we find that H = 0.1615.
If we want to make a model which is just at this limit, we must therefore take the value of the semi-hinge length h = 0.1615. and with these parameters decided, we can arrive at the angles and proportions of the triangles.
They are: a = 1.0133, b = 1.0380, A = 76.64°, B = 85.29°, C = 18.06°.
Once you have cut out and glued together this model you will find that it turns and can be manipulated in a most satisfactory and pleasing manner. The vertices come together and open out widely in a rather dramatic but silky smooth way.

Metamorph 6

When this model is rotated, it turns in a rather curious and much less symmetrical way than any of the others so far. At the critical points, two hinges come together but the third one never joins them. There is always a gap.

It is a fact that the parameters are exactly the same as those for Metamorph 4, namely $\cos\phi = 0.25$, $\phi = 75.5225°$, $e = 1$, $H = 0.5164$. Giving the values: $a = 1.18322$, $b = 1.29099$, $h = 0.5164$, $A = 60°$, $B = 70.8934°$, $C = 49.1066°$.

The gap occurs because although $h = 0.5164$ the reduction of symmetry means that the simple calculation of the maximum hinge length is no longer valid Although the shape of the elements is the same as for Metamorph 4, they have been joined together in a different way and this change produces a metamorph with a distinctly different behaviour.

Here + and - are used to represent mirror images.

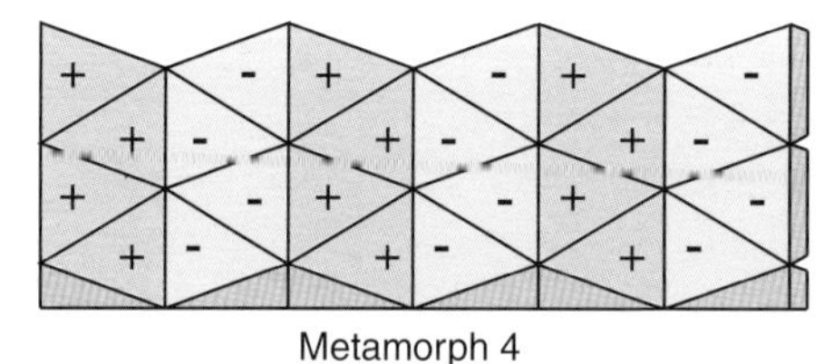

Metamorph 4

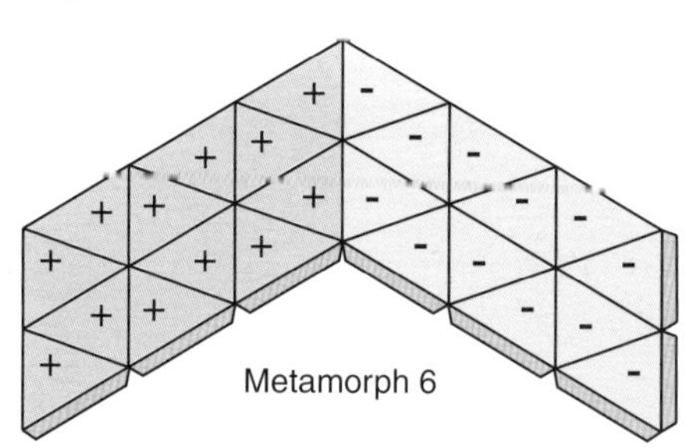

Metamorph 6

You might like to experiment with other combinations and see what happens.

Degrees of Freedom

We now come on to the very important topic of 'degrees of freedom' which is of considerable interest to engineers and mathematicians and has a strong relevance to metamorphs. It really refers to the number of independent pieces of information needed to fix the position of a body in space or within an assemblage.

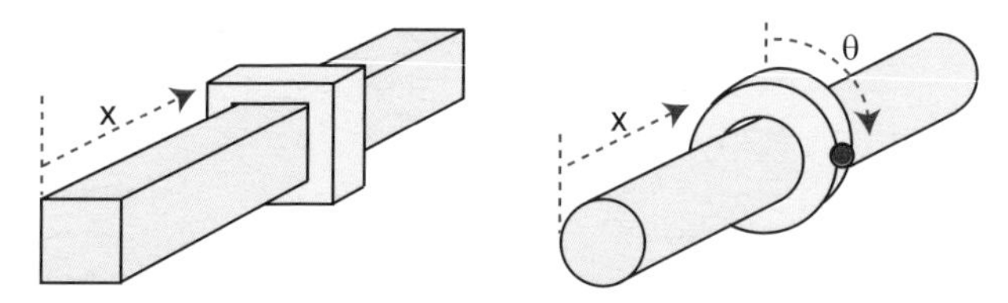

Let us consider first a square ring sliding on a square rod. It can only move backwards and forwards on the rod. It has only one degree of freedom. Another way to look at it is to say that it only requires a single measurement, say its distance from one end to fix its position within the assembly. A circular ring sliding on a circular rod has two degrees of freedom. It is necessary to define how far it is along the rod and at what angle it has been rotated from some base position, say the vertical.

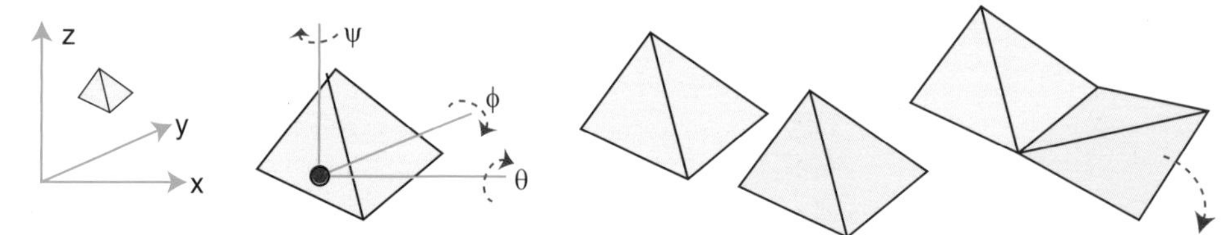

When we talk about a body which can move freely in space, it has six degrees of freedom. A fixed point on the body, say its centre of gravity, needs three spatial coordinates (x, y, z.). Also it can be rotated in three directions, say about the x, y and z axes.

Two independent bodies in space will therefore have twelve degrees of freedom. However, two bodies which are hinged together have only seven degrees of freedom, six to fix the first one and one to define the rotation of the second relative to the first.

A linkage with 7 elements

An assemblage made up of 7 independent elements would have 7 x 6 = 42 degrees of freedom. However, as we have just seen, when a body is hinged to another it loses 5 degrees of freedom. There are 7 hinges and so 35 degrees of freedom are lost, leaving just 7. However, it is the behaviour of the whole assembly we are interested in, not its actual position in space. We must therefore deduct six degrees of freedom for the position in space of the first element. This leaves a single degree of freedom and proves that in general a metamorph which has seven elements will move smoothly along what appears to be a predetermined track.

Metamorph 7

When you have made this model you will observe that it turns in an apparently irregular and unexpected way. It is also immediately apparent that it does really have only one degree of freedom as proved above. Note how it rotates step by step almost as if it is on rails.

All the tetrahedra are still congruent, but of course the arguments about ϕ and the maximum hinge length H applicable to the six-element metamorphs do not apply directly in this case. Observe how the hinges never all come together at the centre but that they do meet in symmetrical pairs during the rotation.
The triangles are all isosceles and the values for the model are:
a = b = 60mm, 2h = 32.2mm, A = B = 74.43°, C = 31.14°

Linkages with 6 and 8 elements

If an assemblage with 6 independent elements is studied in the same way, it has 36 degrees of freedom. Each of its six hinges result in the loss of five degrees of freedom, leaving 6. However one body can be regarded as immobile because we are not interested in its position or orientation in space, just whether the assemblage as a whole will rotate. We therefore must deduct 6 degrees of freedom, leaving the grand total of zero.

This argument seems to prove that an assemblage of six elements has to be immobile! In general, a linkage with six elements will indeed be static and engineers have devoted and are still devoting a great deal of effort to identifying all the possible special linkages with six elements which are mobile. Such linkages are known as 'over-constrained' or 'paradoxical' and information about them is available at www.morpher-magic.com. It lists all the mobile linkages with six or less elements which have been discovered so far. Fortunately all the metamorphs in this book fall into this class of special linkages!

Applying exactly the same argument, step by step to a general rotating ring with n elements produces a value for the number of degrees of freedom of 6n - 5n - 6 = n - 6. If n = 8, the model therefore has two degrees of freedom and will be floppy in use. This confirms in a theoretical way what was observed practically when handling model 2.

Constrained linkages

The mathematics of these special kinds of motion is in general well beyond the scope of this book but certain results are worth quoting, especially those referring to assemblages with six elements. It can be proved that a linkage with six elements will have full-cycle mobility if (but not only if) all six of its hinge lines always intersect one line. The proof makes the general point that mobility is only assured momentarily on any occasion when all hinge lines do intersect one line. In many cases, the smallest movement will ensure that this vital requirement is no longer met and the model will then lose its mobility.

Our six-element models do not have this problem as they have sufficient symmetry to ensure that all the hinge lines intersect the central axis. Look carefully at model 6, the irregular one. Note that its hinges still intersect the central axis and that is why it has full- cycle mobility, even though it does not have the symmetry of the others.

Paul Schatz and his Cube

The study of metamorphs (also generally known as invertible forms) was begun by Paul Schatz (1898 - 1979) who saw in their motion a significance far beyond the interesting fact of full-cycle mobility. Schatz, who was born in Constance, Germany, was a very gifted scholar, who began to study Mathematics, Mechanical Engineering and Philosophy, but shortly before completing his studies, changed over to Astronomy. Then, disenchanted with what he regarded as the abstract approach to Science which he had experienced, he discontinued his scientific studies and enrolled in the Warmbrunn School of Woodcarving. For several years he had his own studio as a sculptor.

Through his discovery in 1929 of the Invertible (or Schatz) Cube, he was inspired to resume his engineering activities with the ideal to 'evolve a technology that would be fully in harmony with the world of nature and the human being'. He later described how aesthetic considerations, as part of an experiential approach to Mathematics, led him to his discovery. He was convinced that the remarkable way in which this cube turns inside-out and continues in the same sense to turn outside-in in an endless rhythm was related to processes in the natural world and regarded this motion as the foundation of an ecological technology. He applied it successfully in mixing technology and water purification.

Schatz also began the study of time forms, which are the lines, surfaces and volumes which are generated by parts of a metamorph during a cycle of motion. One such time form, the oloid, which is generated by a diagonal of the Schatz Cube when one element is held static during a whole cycle is the basis of the innovative and effective Schatz system of water purification.

The Paul Schatz Foundation at www.paul-schatz.ch has the task to organize the extensive scientific and artistic material of his estate, to develop it and make it accessible to all who are interested. It also encourages further research along the lines which he indicated and the development of the potential of numerous ideas he had on new technology. Researchers and practitioners in many disciplines continue his work.

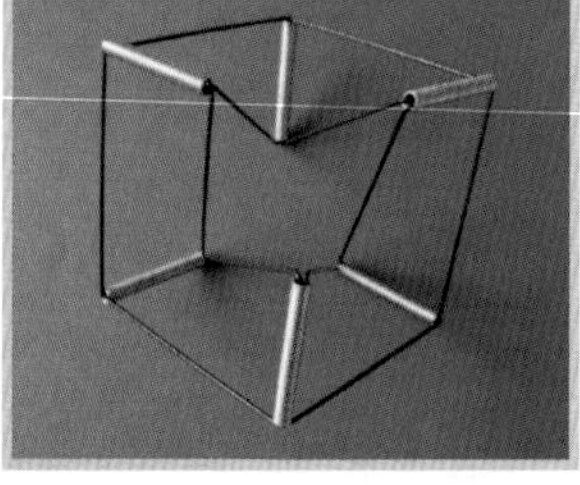

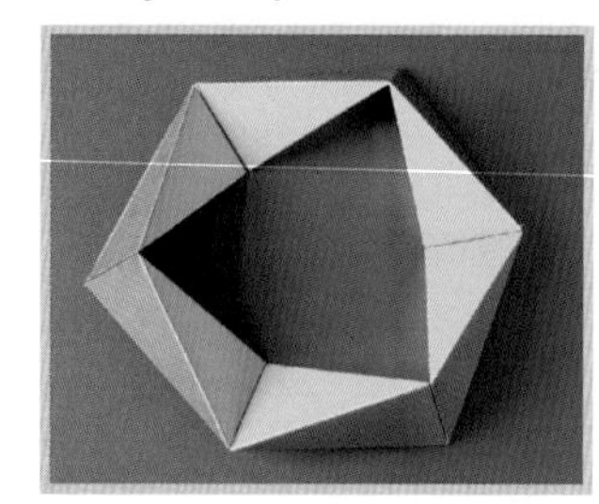

The photographs above show some commercially available toys which are based on his discoveries. The first two are six-element metamorphs, one elegantly made in metal and the other in card. They both have full-cycle mobility and the card one, in blue, fits precisely between two identical immobile fang-shaped pieces to form the Schatz cube. The third item is a wooden 'oloid'.

Metamorph 8

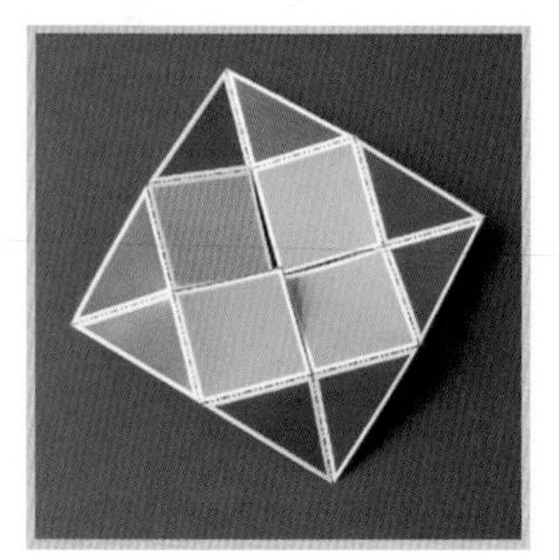

This plane metamorph has full-cycle mobility in a similar sense to the mobility of the earlier models. It turns through its centre in a rather three-dimensional way, even though it is strictly a two-dimensional object.

Observe how four angles of 90° meet at the centre on each side, making a total of 360° and yet the vertices will easily pass through. There is no 'superfluous paper' as in the case of the six regular tetrahedra to obstruct the turning. Note also that the method of manipulation is fundamentally different from that of models 9, 10 and 11.

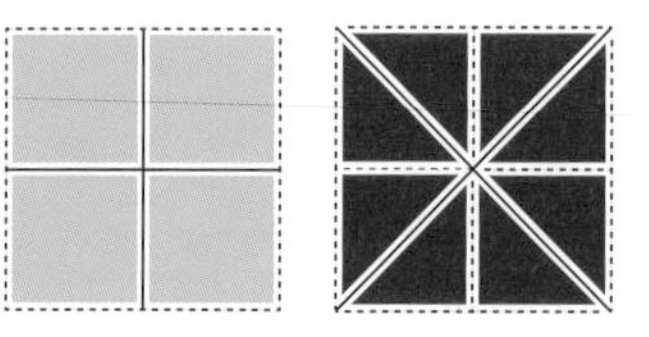

Metamorphs 9 & 10

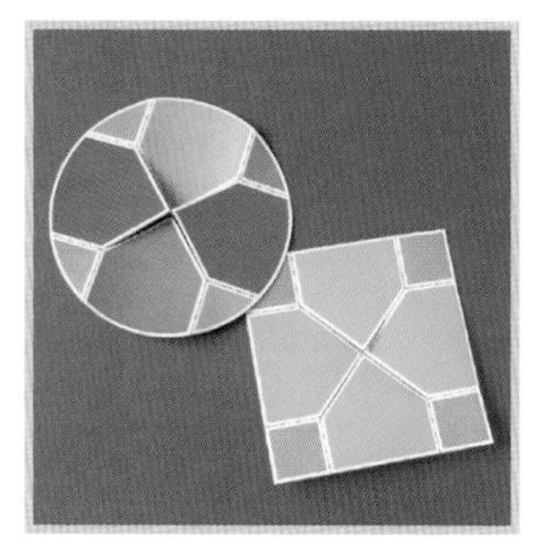

These two models are virtually identical except for their colours and their circular or square outlines. They have full-cycle mobility, but in a rather different form from the previous models. This new kind of mobility requires a sequence of discrete folds in a certain order along several hinges in succession to complete the transformation.

Holding either of the models 'straight' in front of you and using h for the horizontal folds and v for the vertical folds, a simple notation can describe the motion, for example a complete cycle can be described by $h_1\ h_2\ v_1\ v_2\ h_1\ h_2\ v_1\ v_2$.

Observe also that if the order of folds is described by $h_1\ v_1\ h_2\ v_2\ h_1\ v_1\ h_2\ v_2$ the models still have full-cycle mobility but the intermediate shapes are different.

Metamorph 11

This model is different from the three other flat forms because there is only one axis on any single face along which it will open.

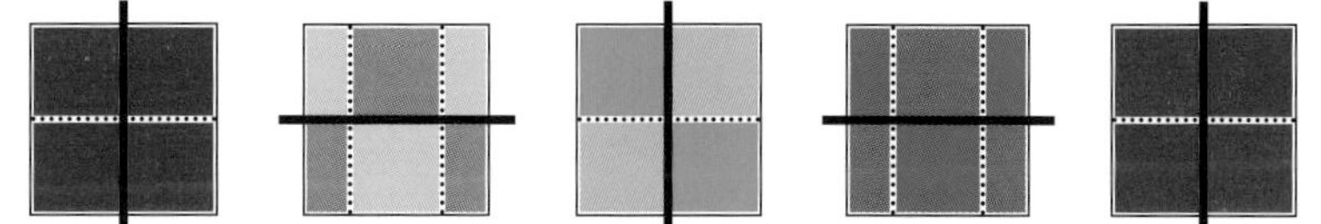

It is convenient to turn this model so that these axes are positioned either horizontally or vertically. They are at right angles and if the first opening happens to be vertical, the second is horizontal and so on alternately until the cycle is complete.

Metamorphs with this pattern of alternate axes at right angles are known as orthomorphs and the remaining three models in the collection are all of this form.

Folding and unfolding cubes

Before considering the final three models, it is well worth thinking about this delightful and well-known toy, made up from eight cubes linked together in pairs in such a way that it can fold, unfold and have full-cycle mobility. It never opens up to leave a hole in the middle and cannot rotate like the metamorphs from 1 to 7. With eight elements and more than two degrees of freedom one might expect it to be floppy. However, the shape of the elements and the clever arrangement of hinges causes it to move in a pleasingly constrained way.

The eight cubes are joined in such a way that the complete model will split along two axes at right angles, although only one at a time. It opens along these axes alternately and so does have full-cycle mobility. For it, we can use the word orthomorph, introduced above.

The photograph above shows two colourful paper models of this type and two of the plastic ones which became so popular in recent years. Not only were units like these used for advertising products and companies but they also served as vehicles for presenting cartoon characters, pithy quotations and even serious information in a novel and dynamic way. Like Metamorph 11 above, its surfaces are flat and square or rectangular, offering convenient spaces for all kinds of commercial images.

Metamorph 12

This model is at first sight an ingenious version of the folding and unfolding cubes models described overleaf. However, the system of hinges is different and it changes shape in a much more dramatic way, opening right out into an open frame. It opens along axes at right angles as all orthomorphs do, either along a single hinge-line or along two parallel ones that have to be operated at the same time. If the model is turned over at any stage in the process, the opening axes are found to be at right angles to those on the other side.

Its dimensions are based on a cube of side six units and the sides of the various squares and rectangles are in the proportions of 6: 3: 2. It can also be seen that the hole in the open frame is of side six units, the same as the initial cube. Some readers might like to make a cube-shaped box of side six units to act as a container for this model.

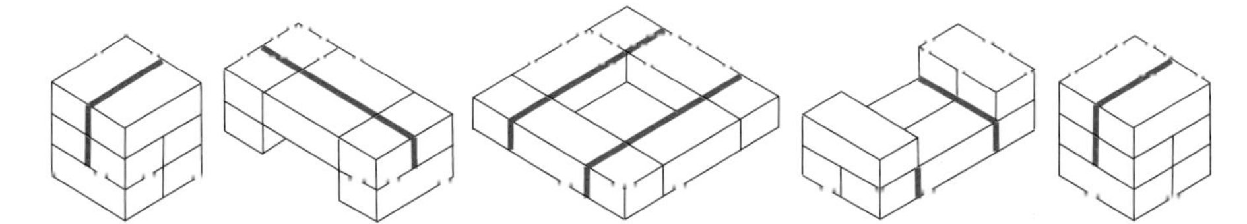

The three distinct 'dividing forms' that the model passes through during a complete rotation can be conveniently described as 'cube', 'office desk', and 'open frame'. The diagrams show where the opening hinge-lines are at each stage and how the orientation of their axes alternate between two directions at right-angles.

The sequence shown in the diagram above show that the metamorph takes a total of four moves to complete the cycle and get back to where it started.

Metamorph 13

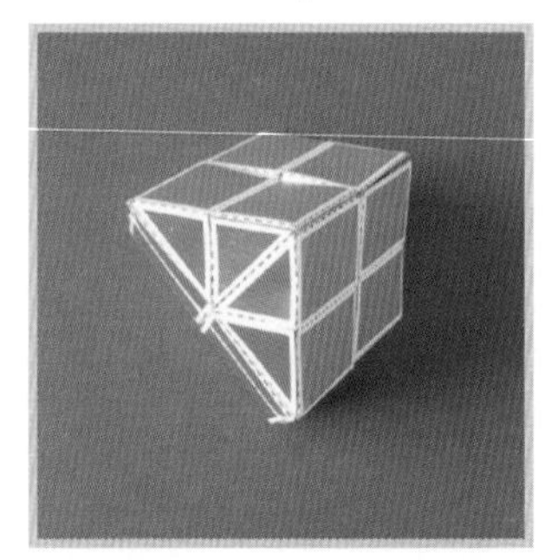

It is very easy to appreciate the beauty of this model as it is turned through its complete cycle. Its four colours appear and disappear as it endlessly cycles between two different forms, which can be called its 'flat form' and its 'dividing form'. However, it will only open along a new axis when it is in the 'dividing form'.

This model is also sometimes known as the 'replicator' because it replicates the same shaped 'dividing form' four times in each cycle. New axes will only open when it is in that one shape. As far as is known, it is the only metamorph which has this unique replicating property. In contrast, the folding and unfolding cubes model has two distinct 'dividing forms' and metamorph 12 has three. If anyone discovers another shape that only has one, please tell the author!

When in its 'dividing form' the sequence of colours through a complete cycle follows the sequence: orange/blue : blue/green : green/red : red/orange or vice versa. When in its 'flat form', each of the four colours is hidden inside in turn.

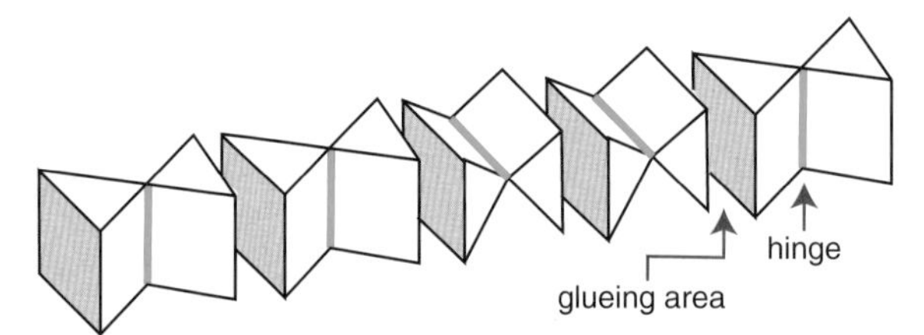

This model is made up of sixteen right-triangle prisms, joined together in a symmetrical but rather cunning way. For the model, each net provides a pair of prisms and the eight pairs are then glued together as indicated by the numbering system to produce the shape that has the required replicating properties.

The sides of the rectangles are also in the proportions of 1:$\sqrt{2}$, as are the right-angled triangles. All the hinges are between pairs of rectangles along their $\sqrt{2}$ sides.

Metamorph 14

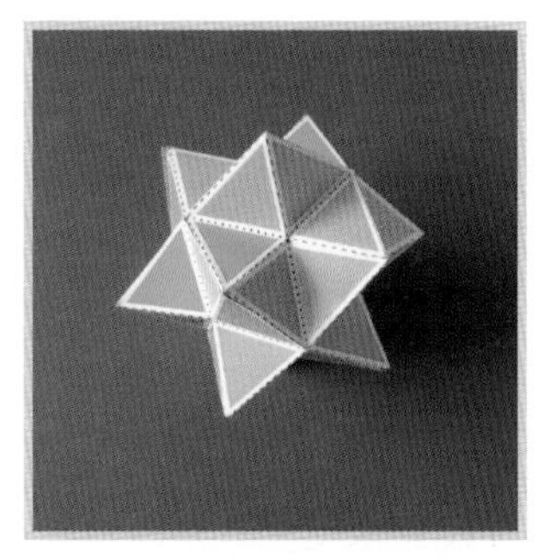

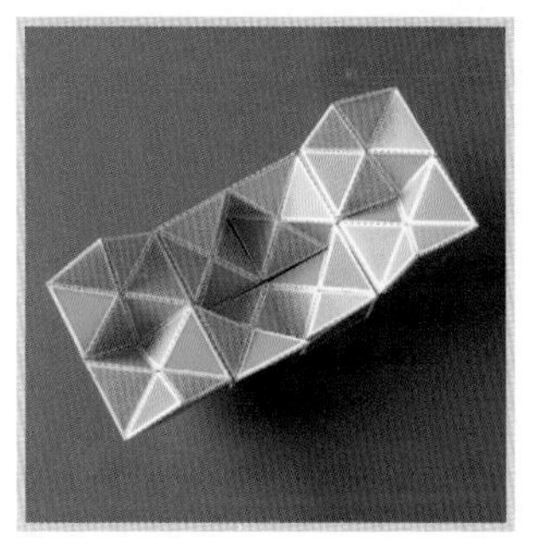

This final model is a splendid example of a metamorph which has 'something extra'. Not only does it have the full-cycle mobility of the eight cube model on page 13, but it has the added attraction of enclosing a 'star-cube' which fits neatly inside. Indeed, with care, it can even be made to go through its full cycle with the star-cube in place.

Although it was probably discovered by Slouthuber and Grafsma, this model is usually known as the 'Yoshimoto Cube', after the Japanese artist and inventor Naoki Yoshimoto who created the splendid gold and silver version which was put on display at the Museum of Modern Art in New York. Cheaper versions in moulded plastic exist, often brightly coloured, under the name of 'Shinsei Mystery'. The angles of the triangles of the star-cube are 54.74°, 54.74° and 70.52° and the sides are in the proportions of $2 : \sqrt{3} : \sqrt{3}$.

Theoretically, the solid star-cube has exactly half the volume of the closed cube and fills exactly half the space inside it. This can be confirmed in a practical way by observing how the cube turns into a similarly sized star-cube at one point in its cycle. In practice, to allow for the thickness of the paper, the solid model has been made very slightly smaller than its theoretical size. Visually this does not make a lot of difference, but it is necessary to avoid damage to the two parts when the model is manipulated.

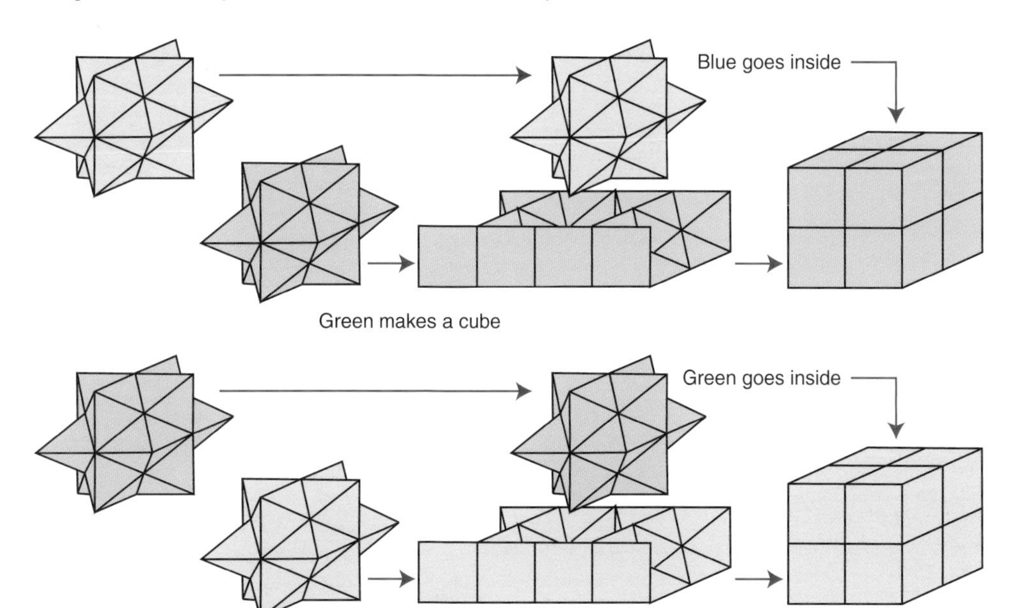

The problem of paper thickness becomes less significant when the model is larger and it is then possible to offer a most surprising demonstration.

If these larger demonstration models are made from thin but strong material it is possible to make two star-cubes of exactly the same size. Each of them is made to the pattern of our outer model and each will open out and fold just like ours. Each is of course a metamorph.

If one is blue inside and out and the other is green inside and out, the blue one in its star-cube form can be placed inside the green one in its cube form, just as you have done with our model. However, now the demonstration can be continued further. The green cube can now be converted into its green star-cube form and the blue star-cube into its cube form. The green then goes inside the blue, just as the blue went inside the green!

As a piece of theatre and as a demonstration of the delights and wonders of mathematics it takes some beating!

Appendix A. Proving the maximum hinge-length formula

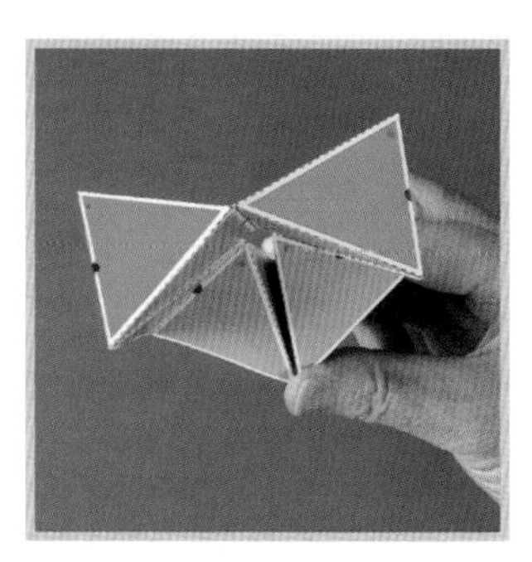

To be able to prove this wonderfully simple and useful formula it is necessary to turn the model in a radically different way, not symmetrically with two hands but in such a way that one hinge is held immobile and horizontal. Keep the model symmetrical about the vertical plane through the fixed hinge. Notice how it seems to climb around the fixed hinge. At some stage in the process you need to change hands.

The proof considers two positions of a single tetrahedral element with one of its hinges held horizontal in just that way. As before, H represents the maximum semi-hinge length.

It helps to take one end of the fixed hinge as the origin and to regard the hinge direction as the x axis. The coordinates of A, the mid-point of the hinge are therefore (H, 0, 0). The far end of the hinge B has coordinates (2H, 0, 0)

The first position to consider is the one where the elemental length is also horizontal, this time parallel to the y axis. It is best to work with unit lengths and so the coordinates of C, the other end of the elemental length, are (H, 1, 0). The point D is where the continuation of the move-able hinge meets the yz-plane. Since ϕ is the angle of twist between neighbouring hinge lengths, the coordinates of D are (0, 1, $H\tan\phi$). Now we consider the rotation of the element about the horizontal hinge OD to bring the model as a whole into a state of three-fold symmetry. It will take a rotation of an unknown angle α to do this. Because of the symmetry, C will arrive at a point E which lies in the plane y = xtan 60°.

This is one of the four occasions when three alternate hinges are horizontal. If the hinge lengths are at their maximum they also meet on the central axis, leaving no space at the centre.

The vertical from E lies in this plane and meets the xy-plane at F (H ,Htan60°, 0). The rotation of α about the x-axis brings the point D to F on the z-axis. By alternate angles, we can see that <ODG is also α.

3-fold symmetry

Now that we have two triangles ODG and AEF, each with an angle α and a right angle we can see that they are similar. Their sides are therefore in proportion. Writing them in table form to be sure that the sides are in the right order we have -

Triangle	Opposite α	Opposite 90 - α	Hypotenuse
❶ ODG	1	$H\tan\phi$	$\operatorname{cosec}\alpha$
❷ AEF	$\sin\alpha$	Htan60	1
❶ x $\cot\phi$	$\cot\phi$	H	$\cot\phi\operatorname{cosec}\alpha$
❷ x cot60°	cot 60 $\sin\alpha$	H	cot 60°

Since both of these transformed triangles now have a common length H, their hypotenuses and the sides opposite α must also be equal. From each pair of alternatives, we can select the expression which does not contain α, namely $\cot\phi$ and cot60°. Since this triangle is right-angled we can apply the 'Theorem of Pythagoras' to it.

$$\cot^2\phi + H^2 = \cot^2 60° \text{ and hence } H = \sqrt{\cot^2 60° - \cot^2\phi}$$

It is easy to see how a similar argument can be applied to a ring with 2n elements.

$$\text{Hence } H = \sqrt{\cot^2 \frac{180°}{n} - \cot^2\phi}$$

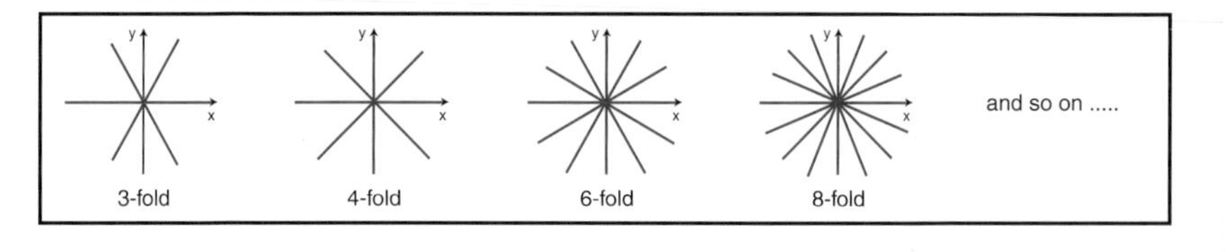

METAMORPH 1

Make this symmetrical diamorph in the form of a rotating ring of fourteen regular tetrahedra.

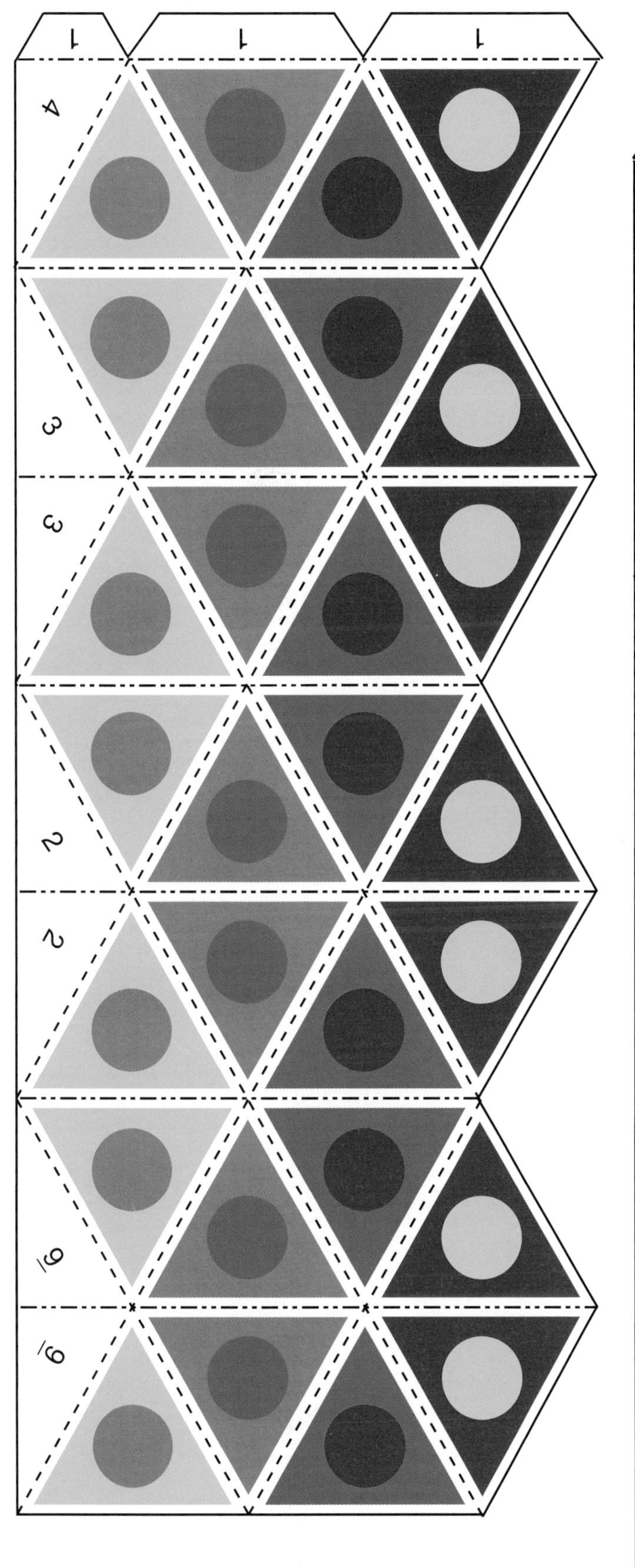

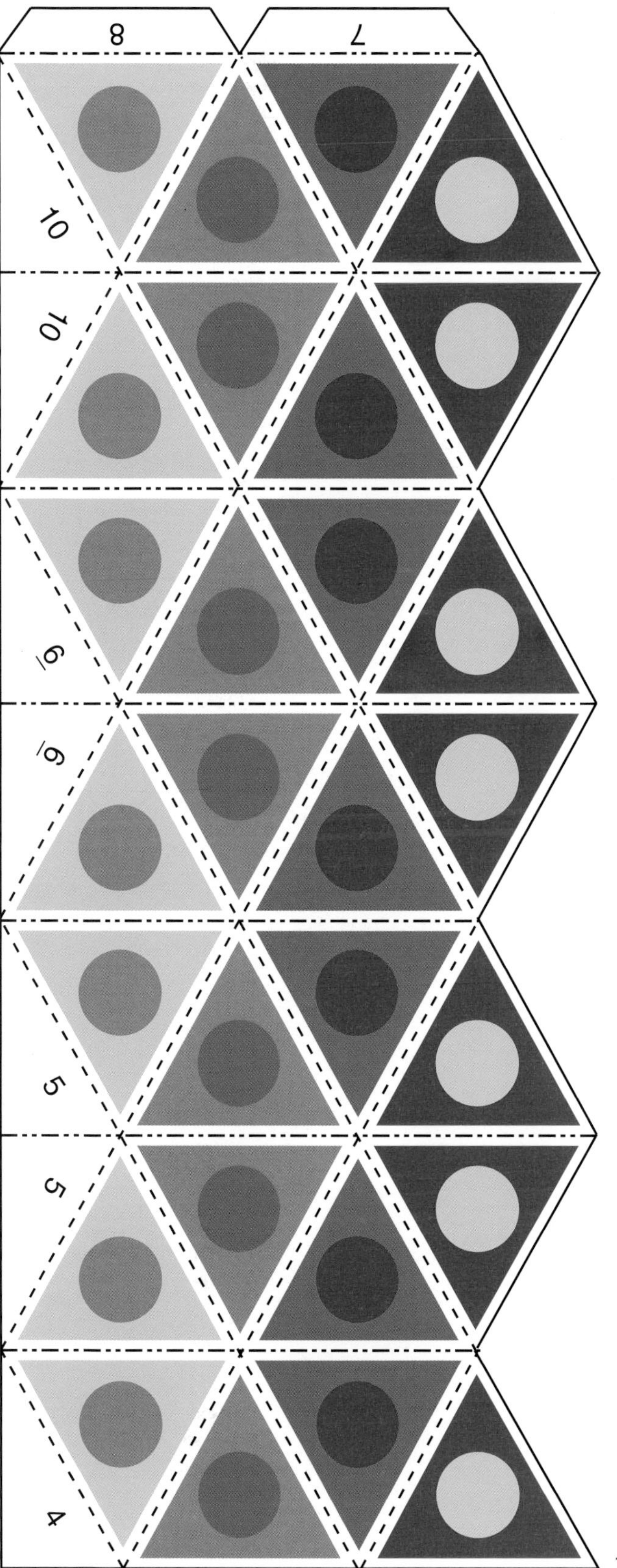

4
10
3
10
3
6
2
6
2
5
9
5
9
7
8
4
1
1
1

METAMORPH 2

Make this symmetrical diamorph in the form of eight linked regular tetrahedra, the smallest number of regular tetrahedra that have full-cycle mobility.

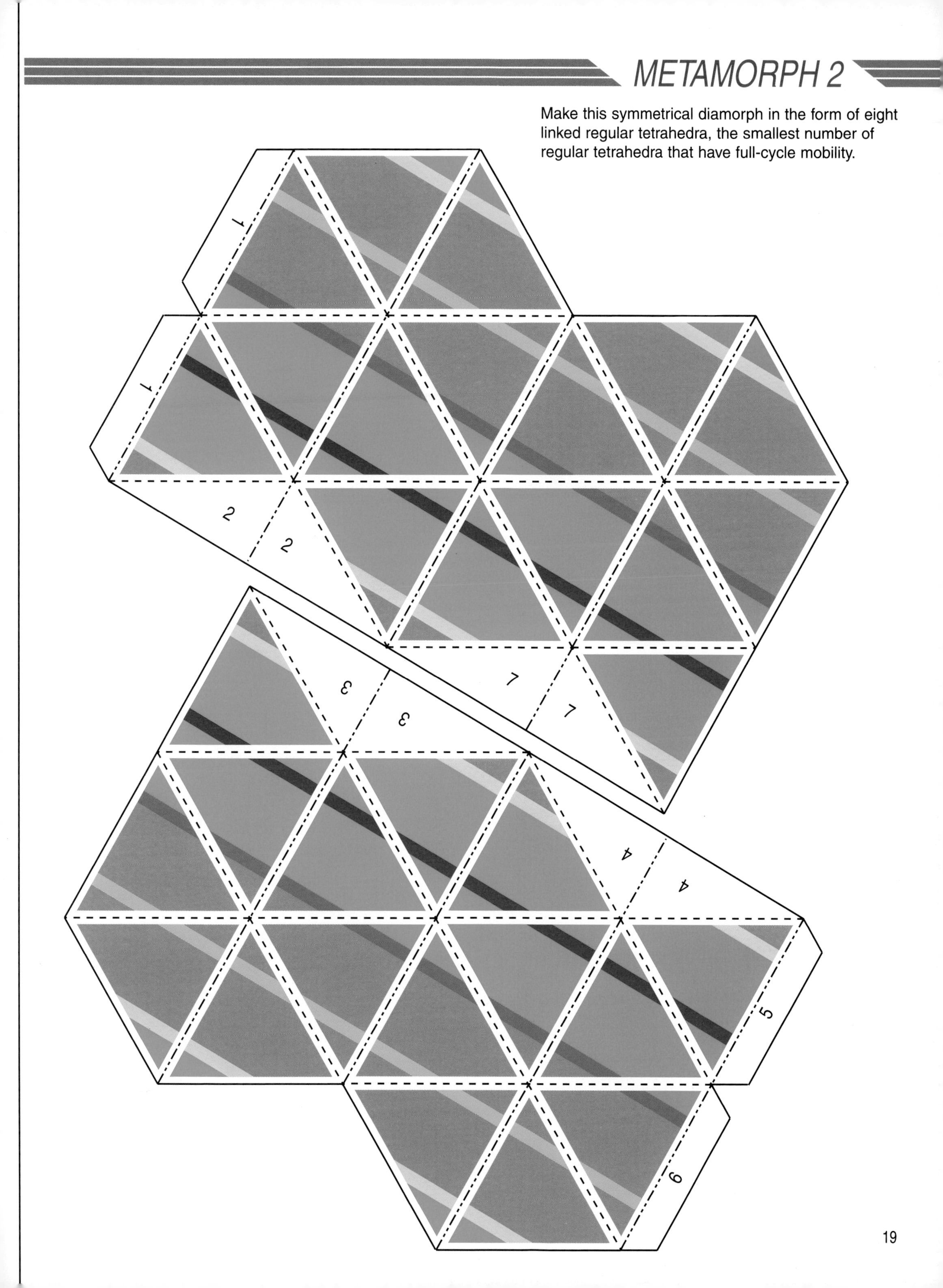

METAMORPH 3

Make this symmetrical diamorph in the form of a classic six-element kaleidocycle. Six faces have been removed to give it a flower-like aspect.
Having glued flaps 1, 2 & 3, rotate through the centre until the green faces are completely inside. Then spread a thin line of glue along each of the pale yellow lines to join the pairs of creases together.

1

2

3

3
1
2

METAMORPH 4

Make this six-element symmetrical diamorph using specially calculated scalene triangular faces. The result is a metamorph that forms plane surfaces four times in each complete cycle.

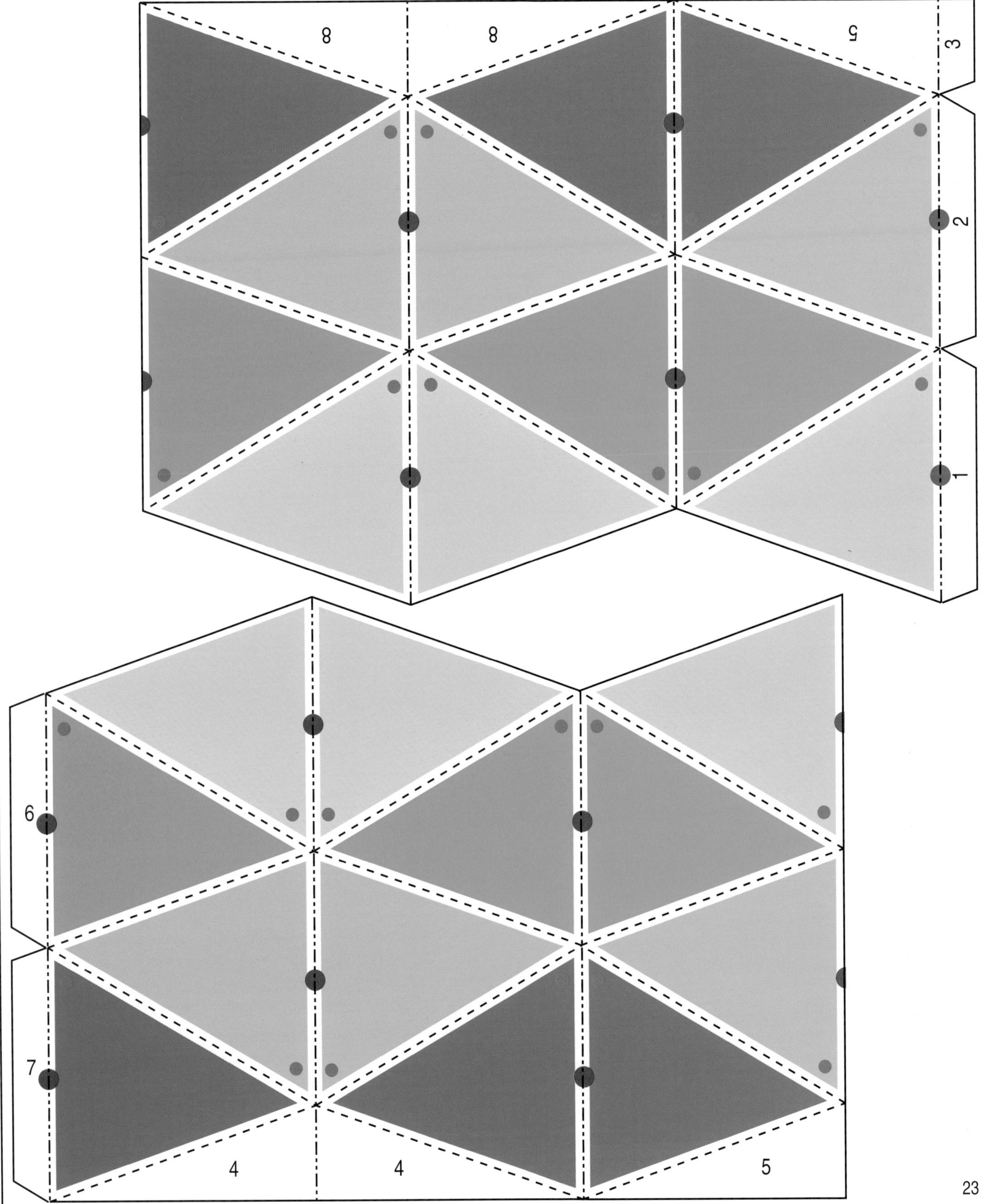

7
6
5
8
8
4
4
5
1
2
3

METAMORPH 5

Make this six-element symmetrical diamorph where the angle of 'twist' is near to its lowest possible value. In this limiting case, the model opens out and closes in a most attractive fashion.

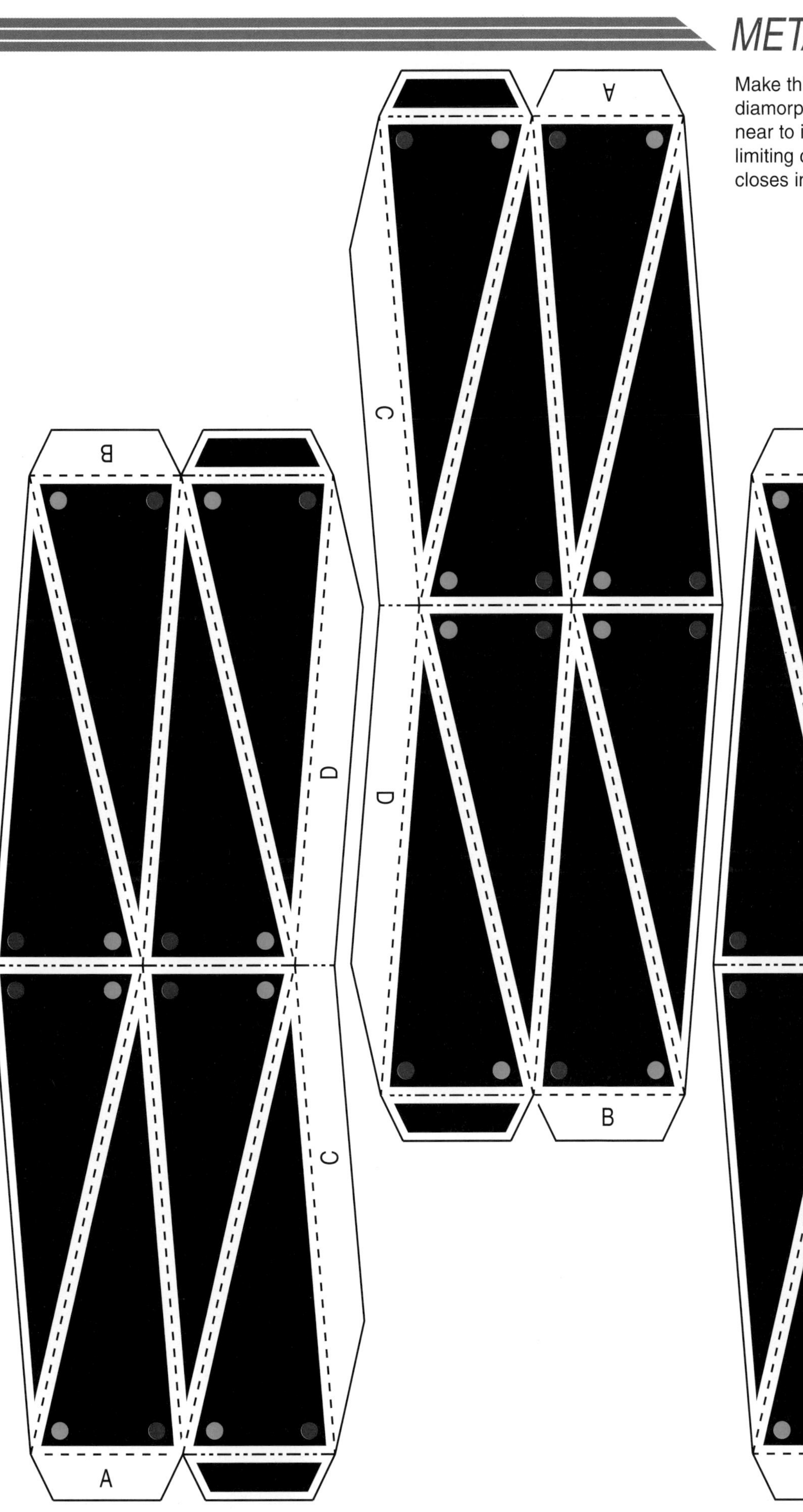

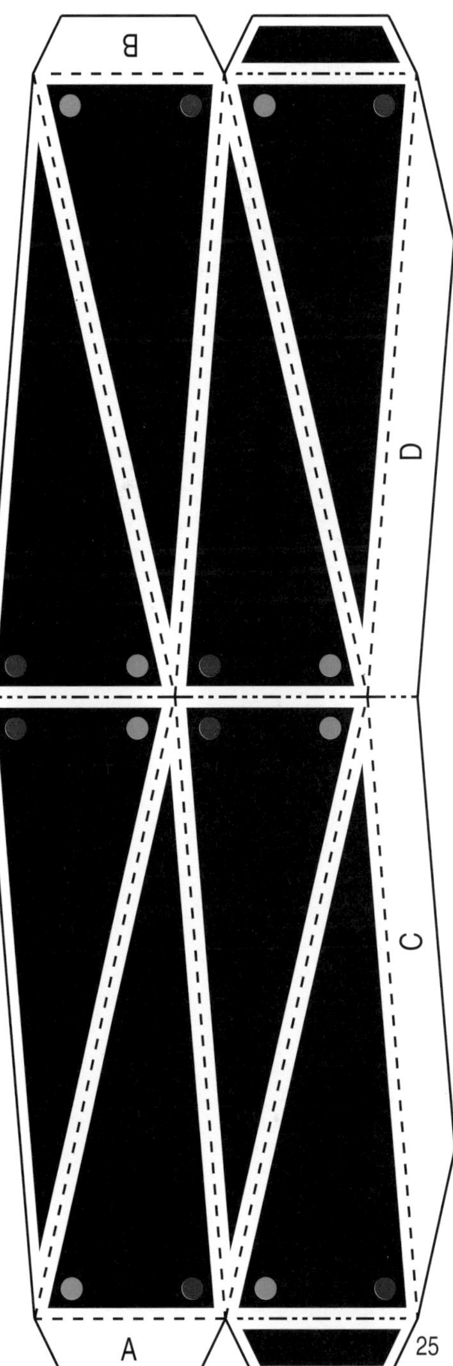

A 2

C

B 2

D 1

D

D

D

B 3

C

C

A 3

A 1

METAMORPH 6

Make this asymmetrical diamorph of six-elements where pairs of surfaces come together to produce an irregular and unexpected motion.

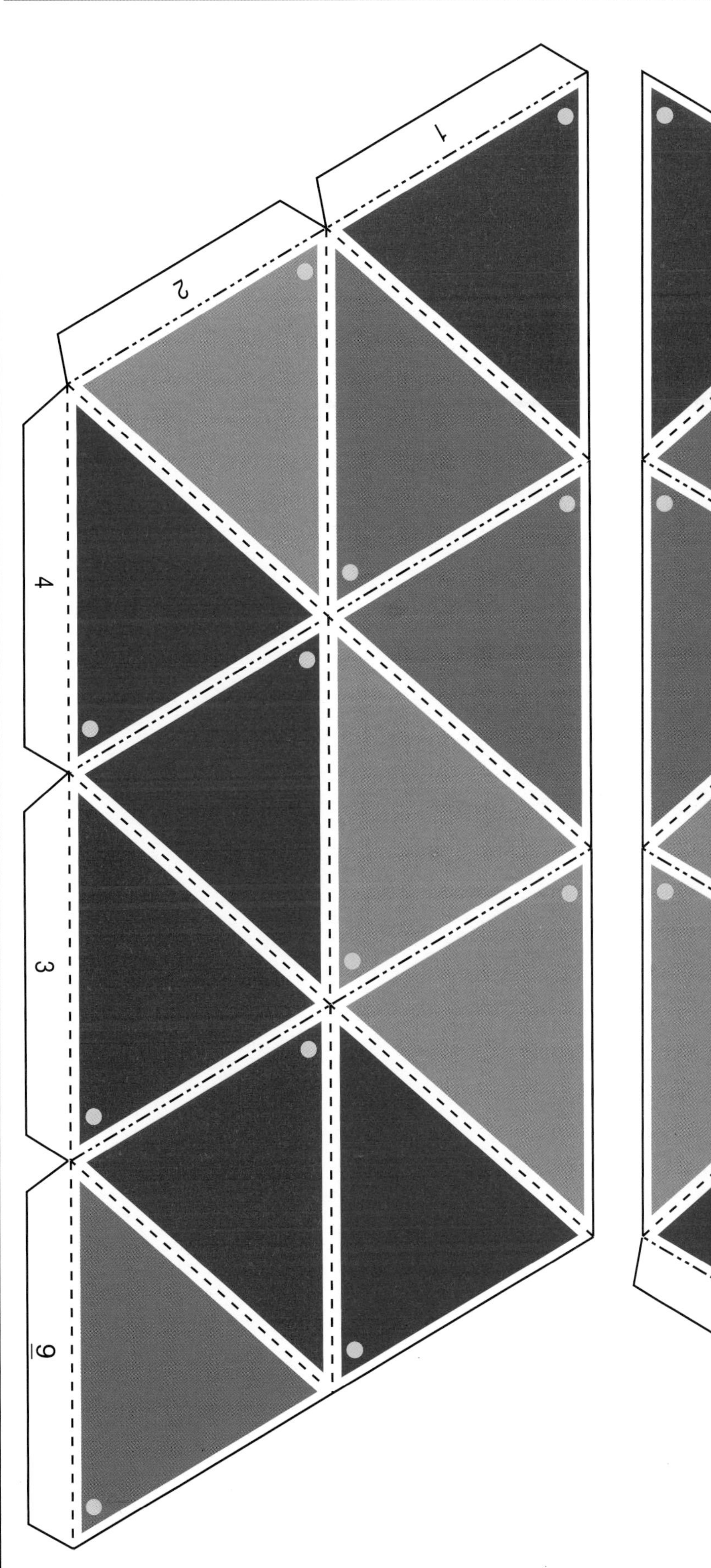

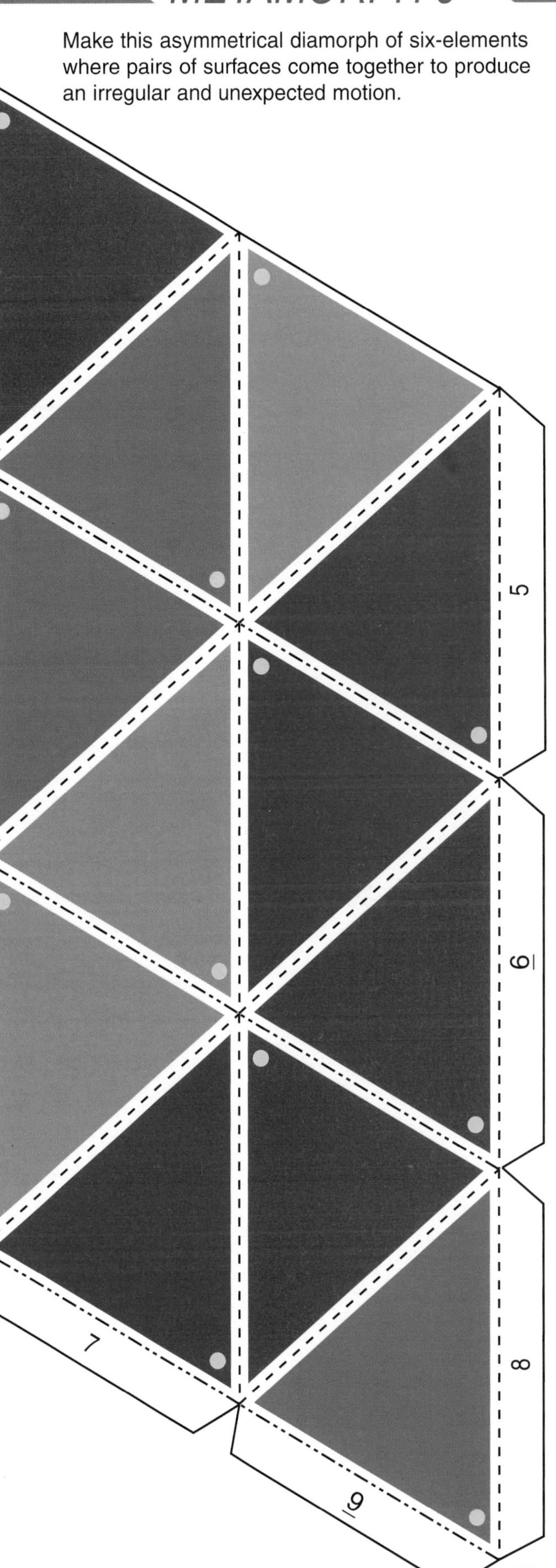

METAMORPH 7

Make this seven-element asymmetrical diamorph with just a single degree of freedom. This causes it to turn rather strangely and smoothly, almost as if it is moving on invisible rails.

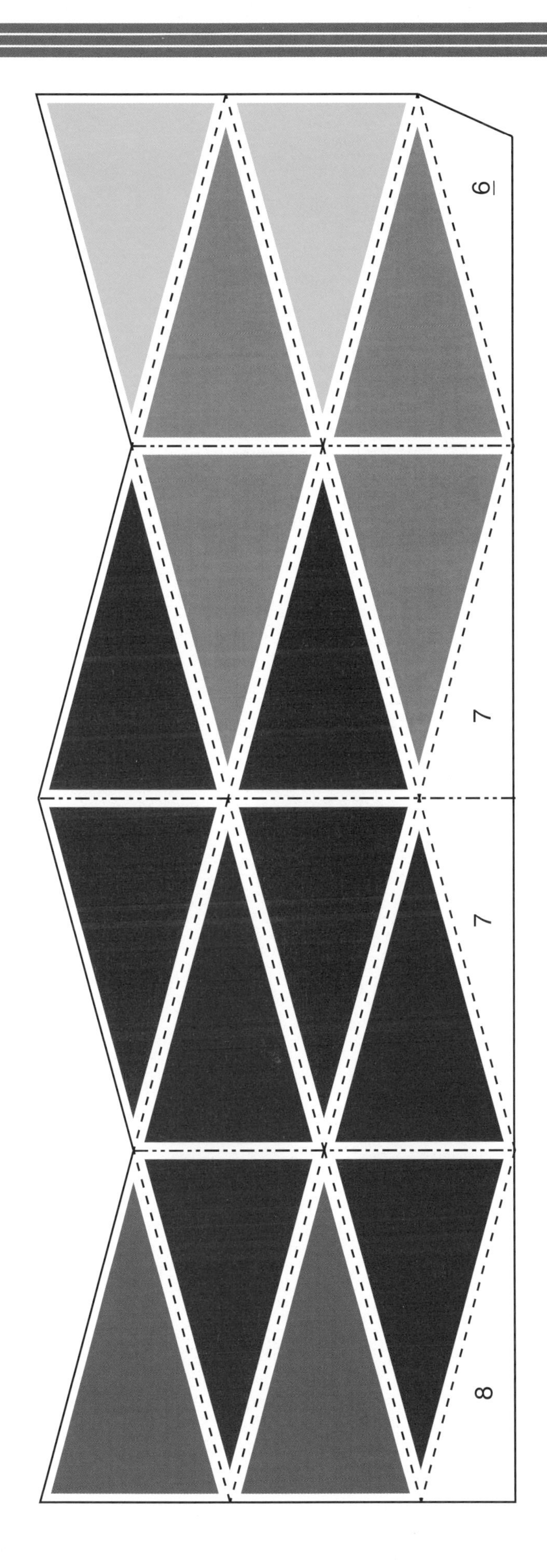

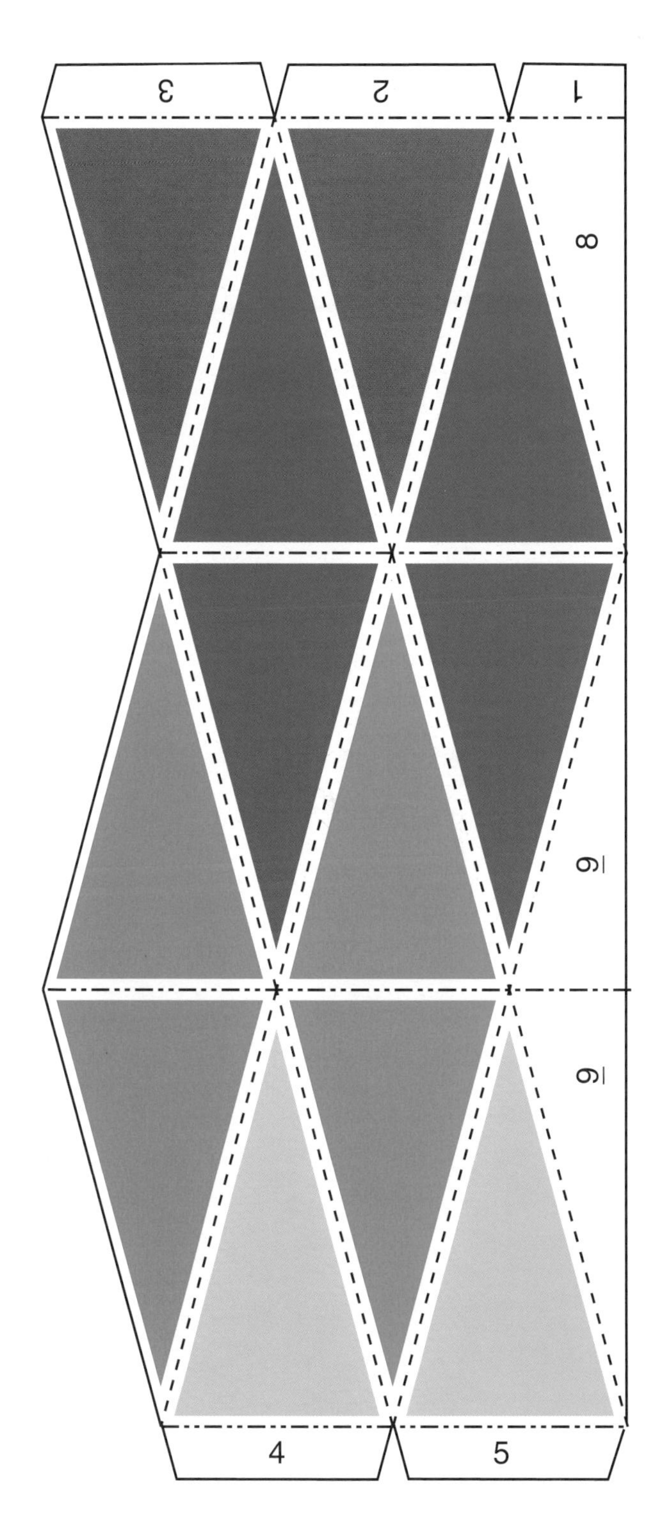

5 4

9

8

7

9 7

9

8

1 2 3

Make these three square plane metamorphs and see how they move, fold and change their shapes in distinctly different ways. Note that there are internal cuts to make with a scalpel.

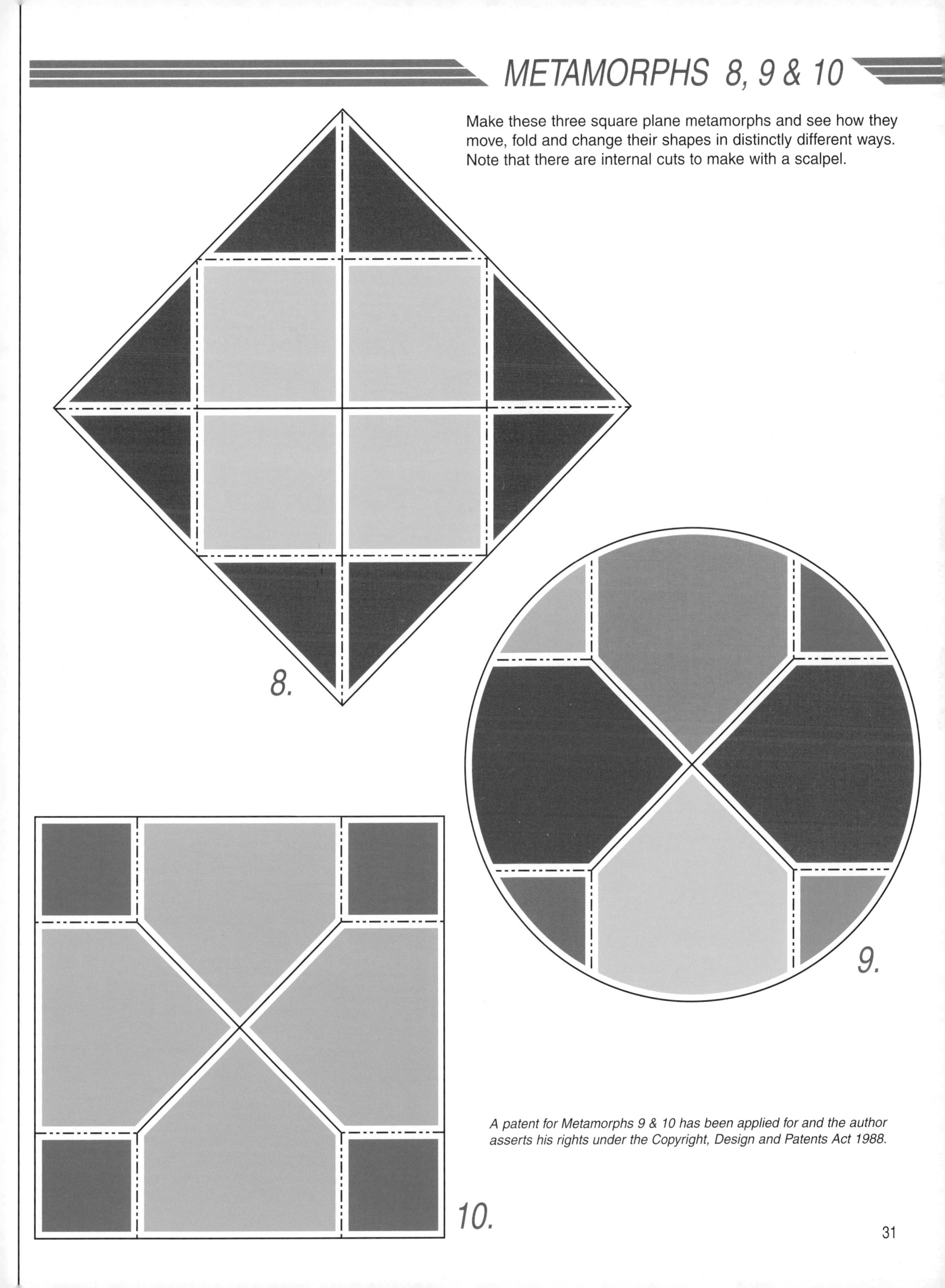

A patent for Metamorphs 9 & 10 has been applied for and the author asserts his rights under the Copyright, Design and Patents Act 1988.

METAMORPH 11

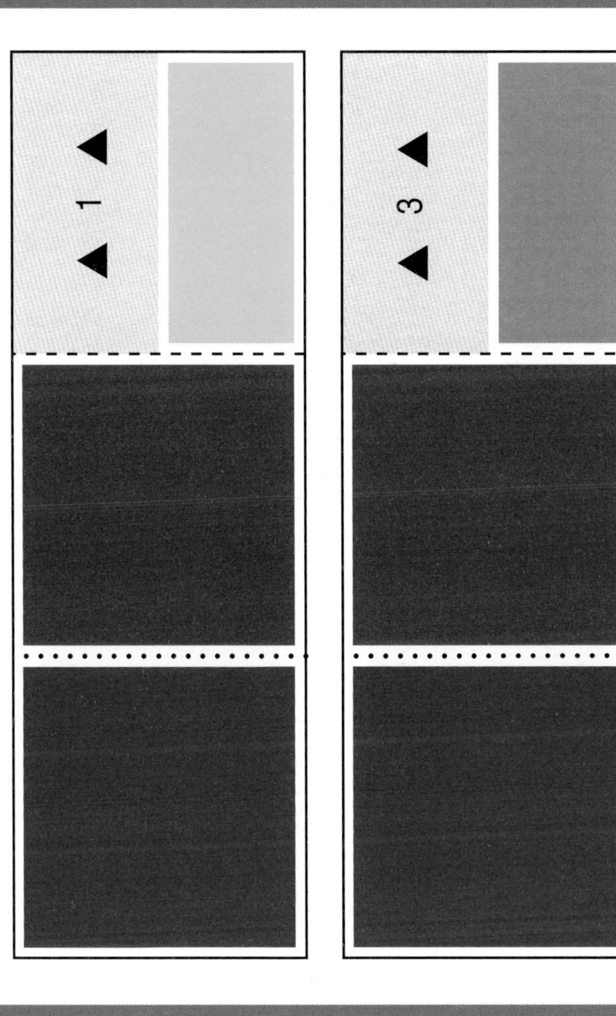

Make this simple plane orthomorph which illustrates in two dimensions some of the folding and unfolding properties of three-dimensional orthomorphs.

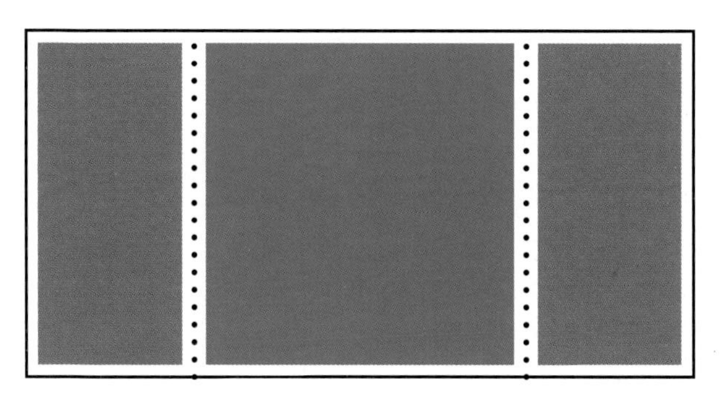

A patent for Metamorph 11 has been applied for and the author asserts his rights under the Copyright, Design and Patents Act 1988.

METAMORPH 12

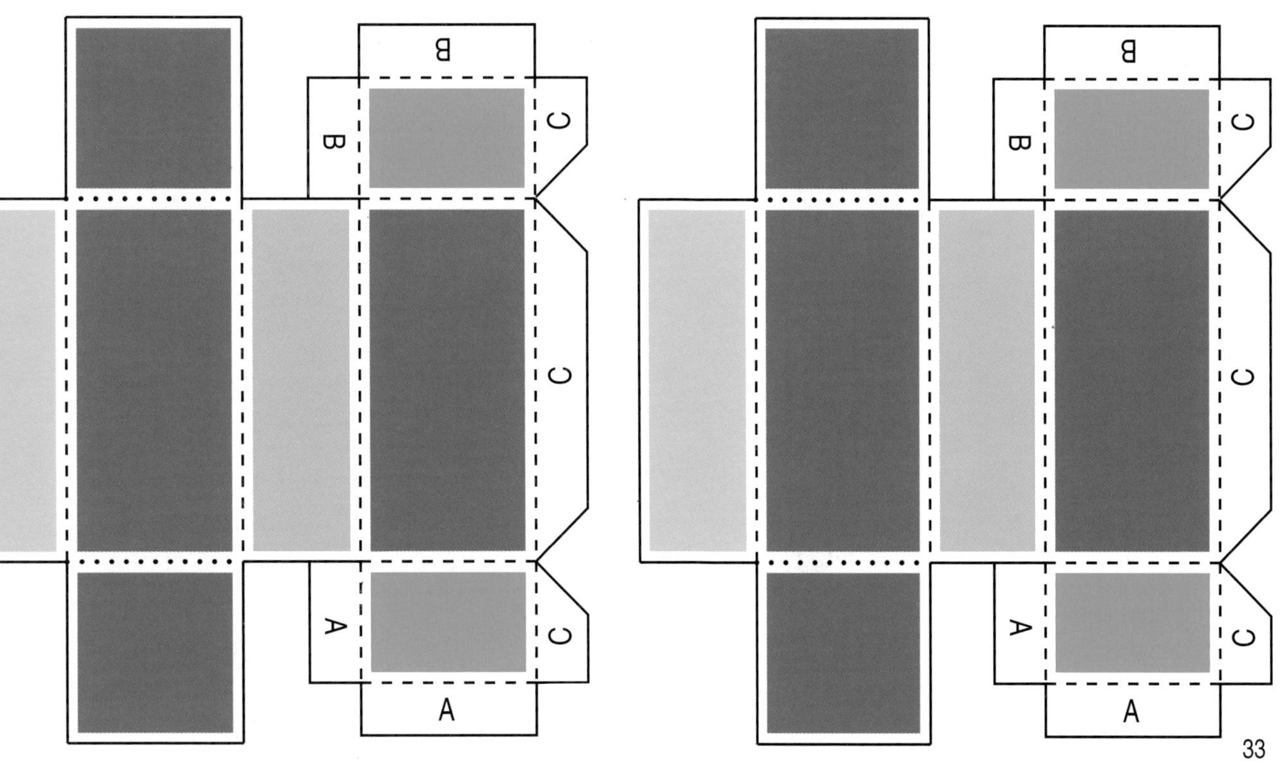

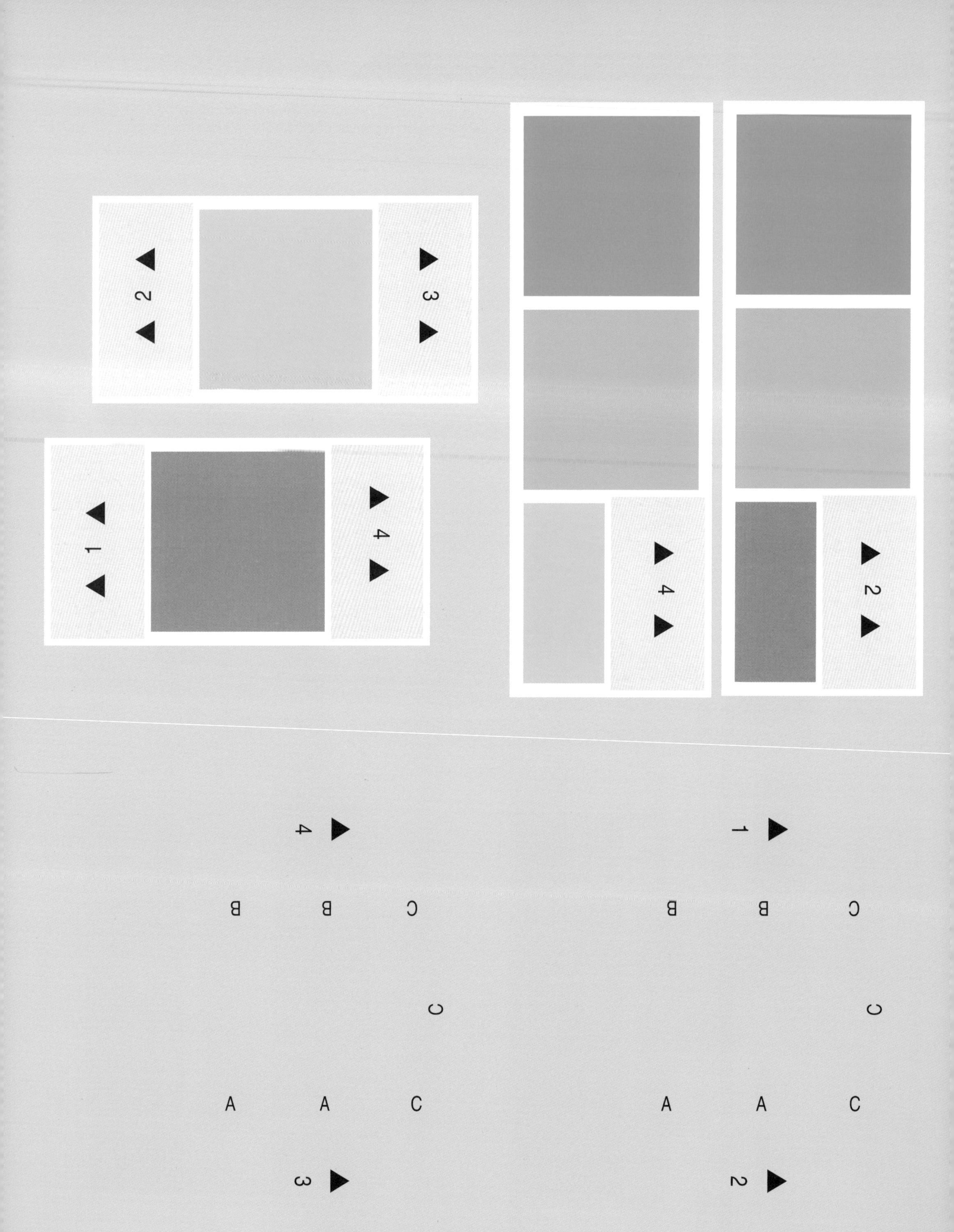

2
3
1
4
4
2
4
1
C B B
C B B
C
C
C A A
C A A
3
2

Make this three-dimensional orthomorph which during its cycle changes its shape dramatically between a simple cube and a rather surprising open frame.

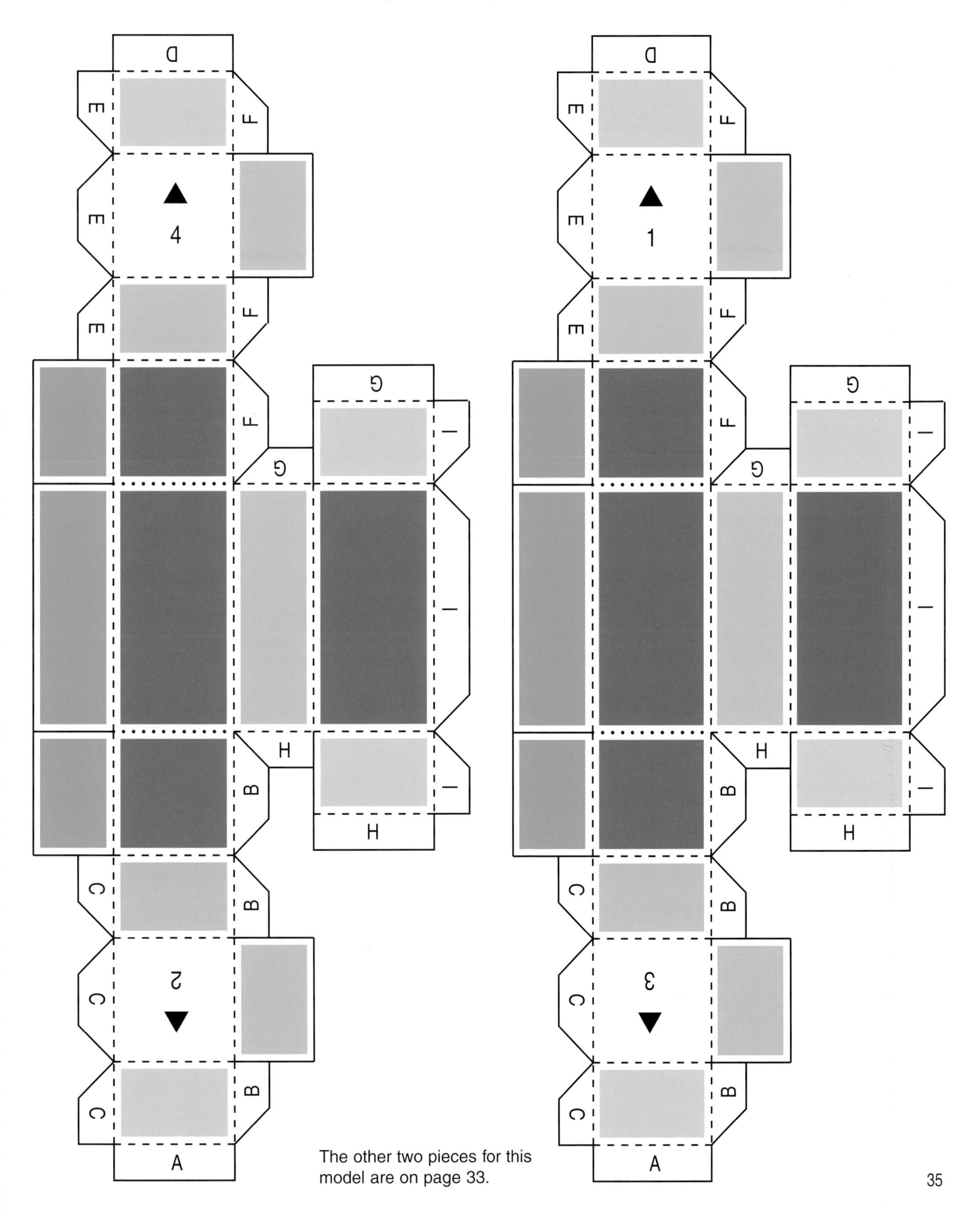

The other two pieces for this model are on page 33.

F
F
F
E
E
G
D E
G I
I
H I
A C
H
C
C
B
B
B

F
F
F
E
E
G
D E
G I
I
H I
A C
H
C
C
B
B
B

METAMORPH 13

Make this unique orthomorph of sixteen triangular prisms joined together in a cunning way which returns to the same shape at each 'dividing' stage of the cycle.

The eight pieces for this model are on pages 37 & 39.

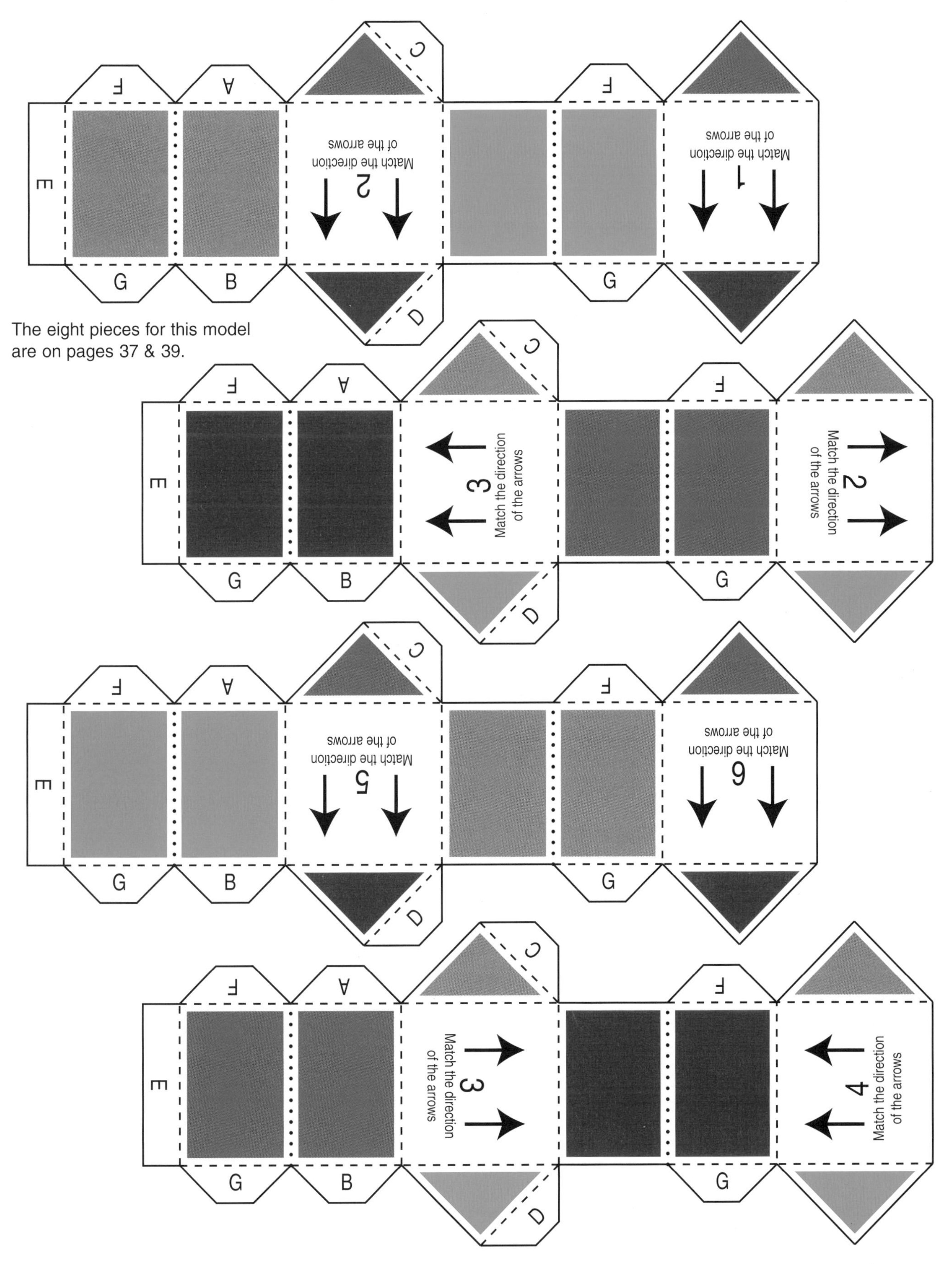

F F A

C

E

D

G G B

F F A

C

E

D

G G B

F F A

C

E

D

G G B

F F A

C

E

D

G G B

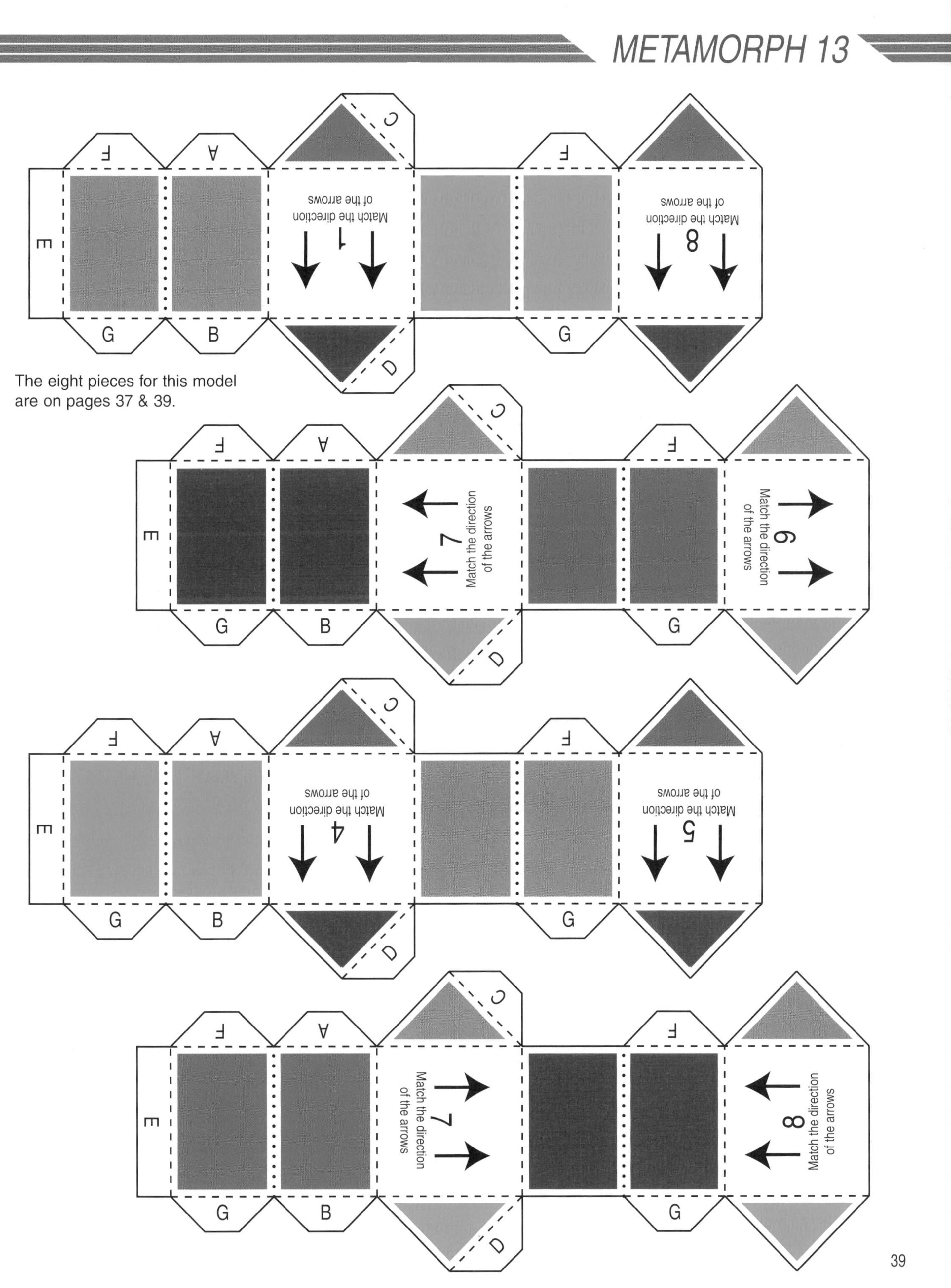

The eight pieces for this model are on pages 37 & 39.

F F A

C

E

D

G G B

F F A

C

E

D

G G B

F F A

C

E

D

G G B

F F A

C

E

D

G G B

METAMORPH 14

Make this striking orthomorph with a hollow centre which exactly encloses an unchanging star-cube. Together they produce some beautiful and remarkable combinations of shapes.

The pieces for this model are on pages 41, 43, 45 & 47.

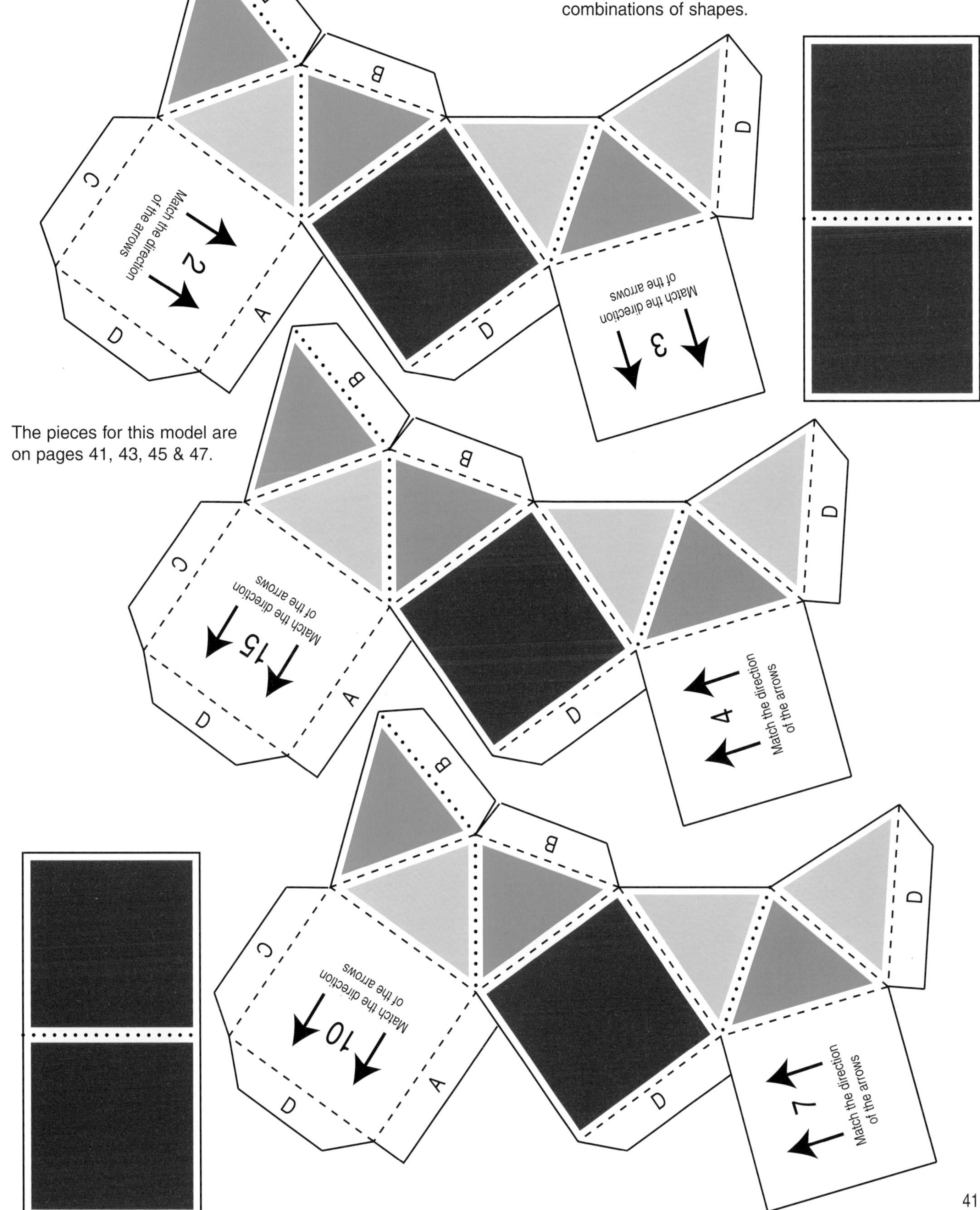

1

Match the direction of the arrows

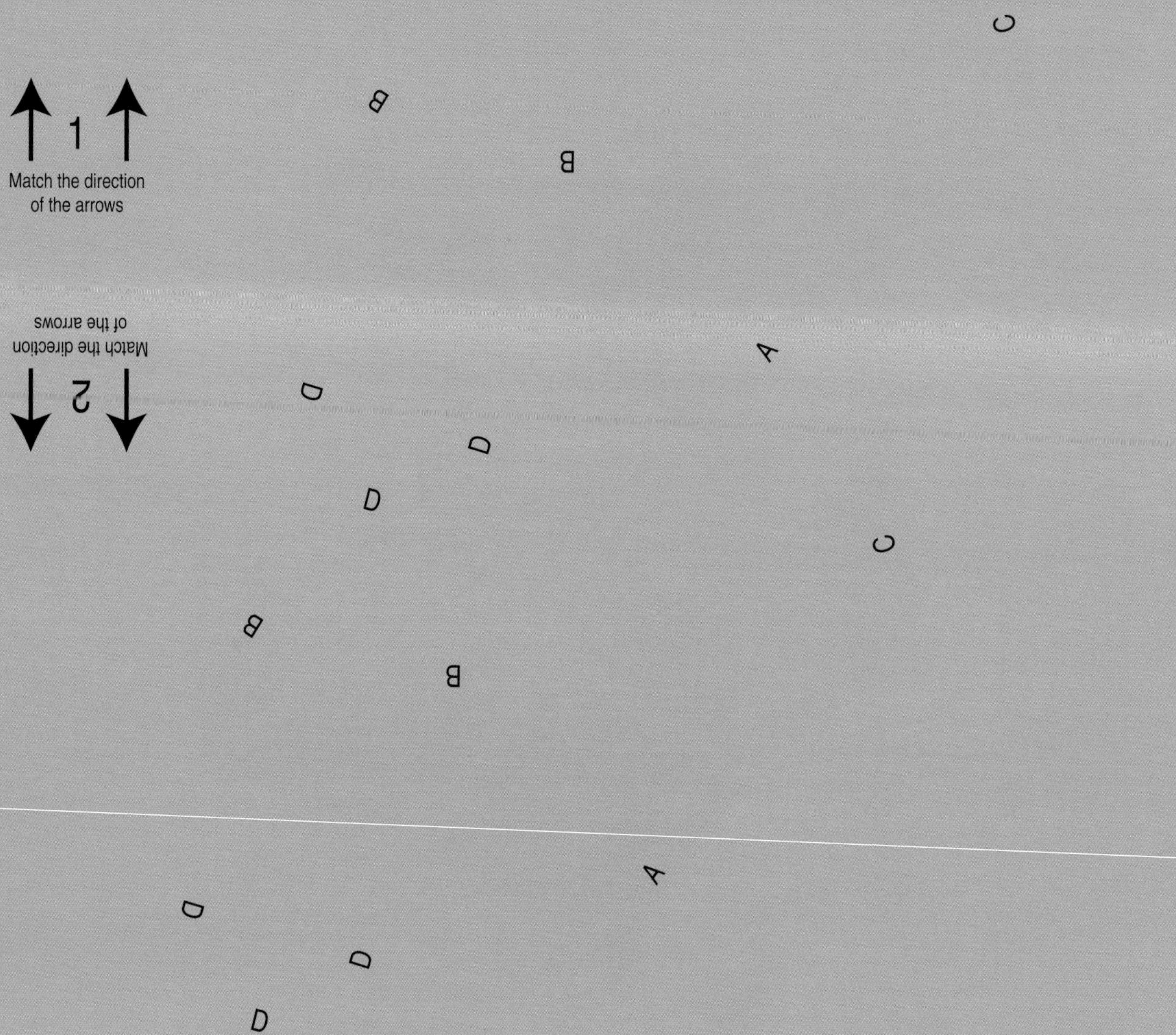

2

Match the direction of the arrows

3

Match the direction of the arrows

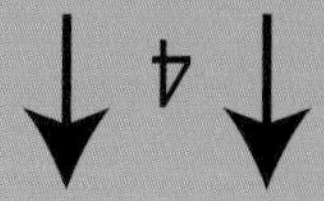

4

Match the direction of the arrows

METAMORPH 14

The pieces for this model are on pages 41, 43, 45 & 47.

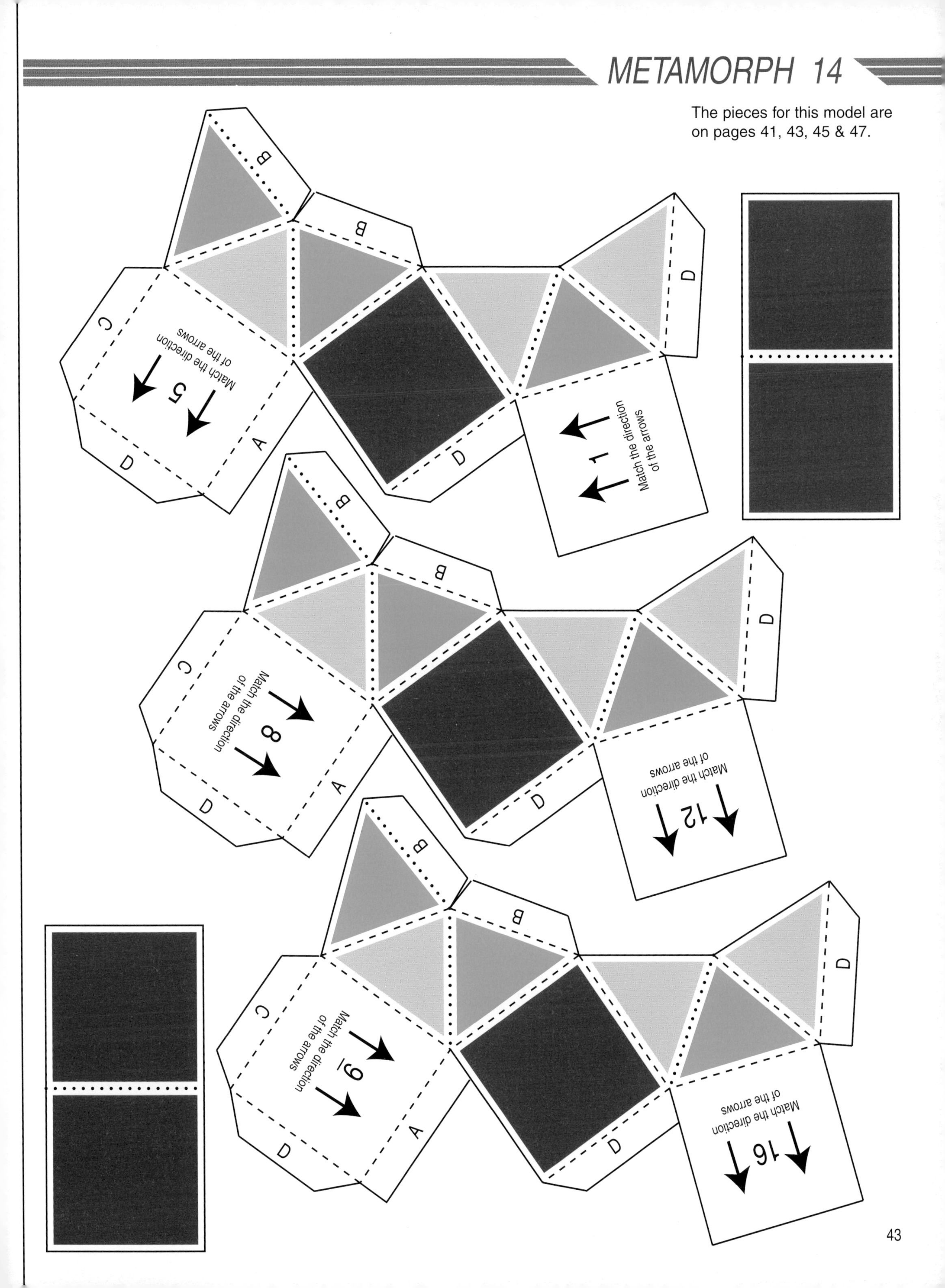

5

Match the direction of the arrows

6

Match the direction of the arrows

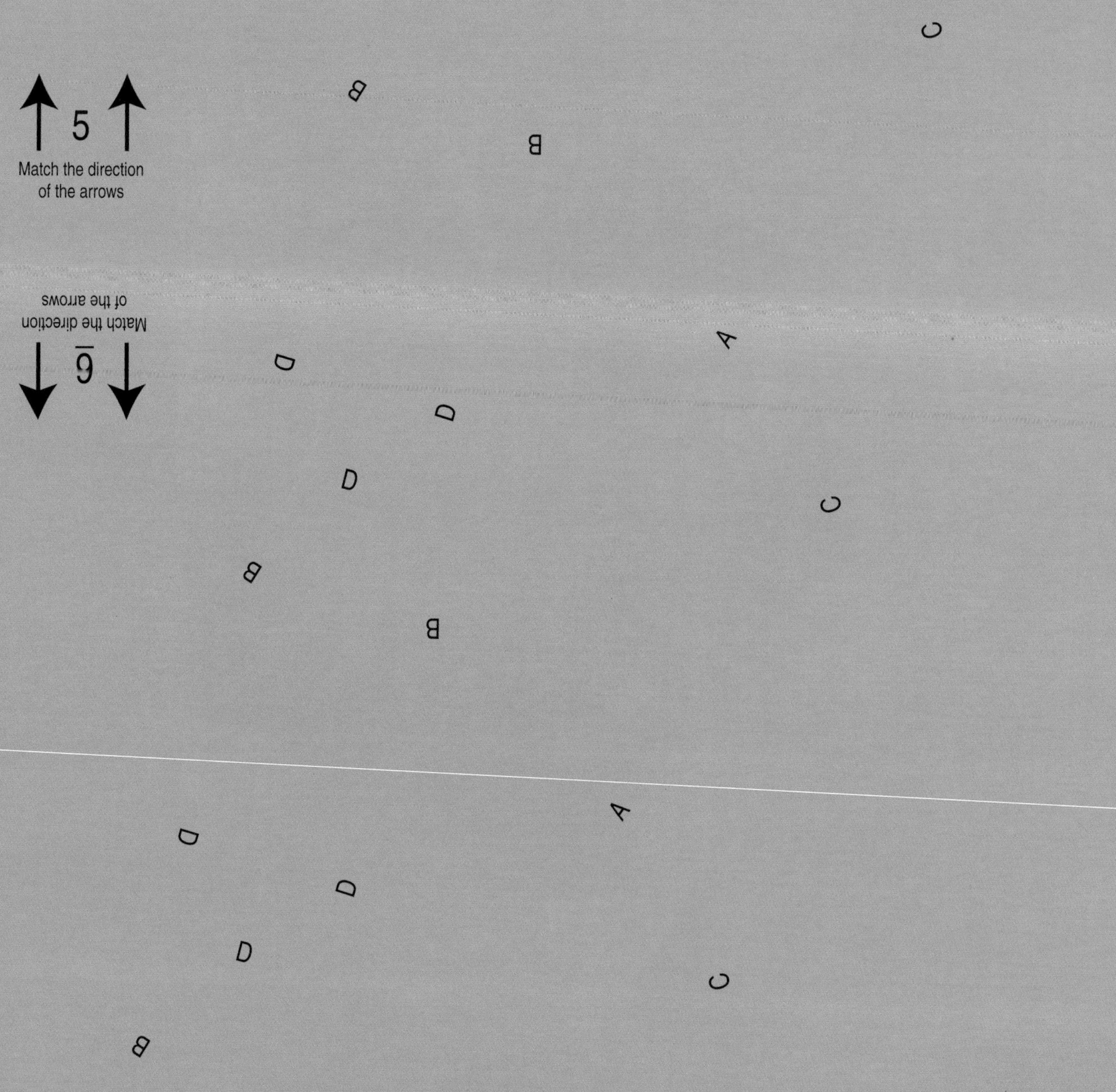

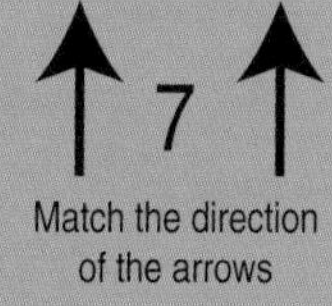

7

Match the direction of the arrows

8

Match the direction of the arrows

METAMORPH 14

The pieces for this model are on pages 41, 43, 45 & 47.

B

B

D

C

Match the direction of the arrows

6

A

D

D

Match the direction of the arrows

13

B

B

D

C

Match the direction of the arrows

14

A

D

D

11

Match the direction of the arrows

C

B

B

A

D

D

D

C

B

B

A

D

D

D

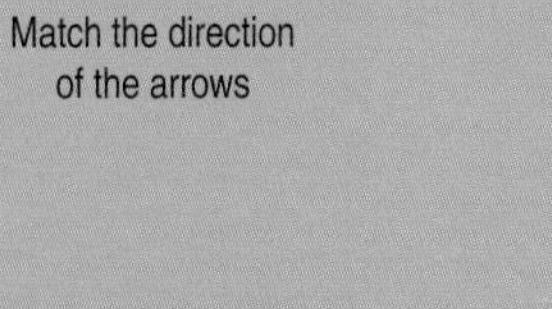

15
Match the direction of the arrows

13
Match the direction of the arrows

11
Match the direction of the arrows

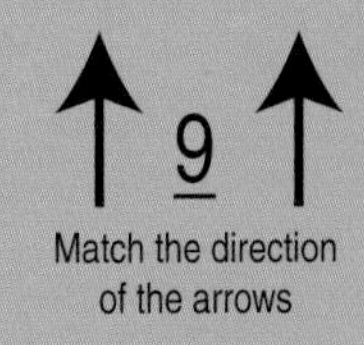

9
Match the direction of the arrows

16
Match the direction of the arrows

14
Match the direction of the arrows

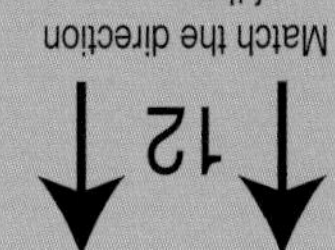

12
Match the direction of the arrows

10
Match the direction of the arrows

METAMORPH 14

The pieces for this model are on pages 41, 43, 45 & 47.

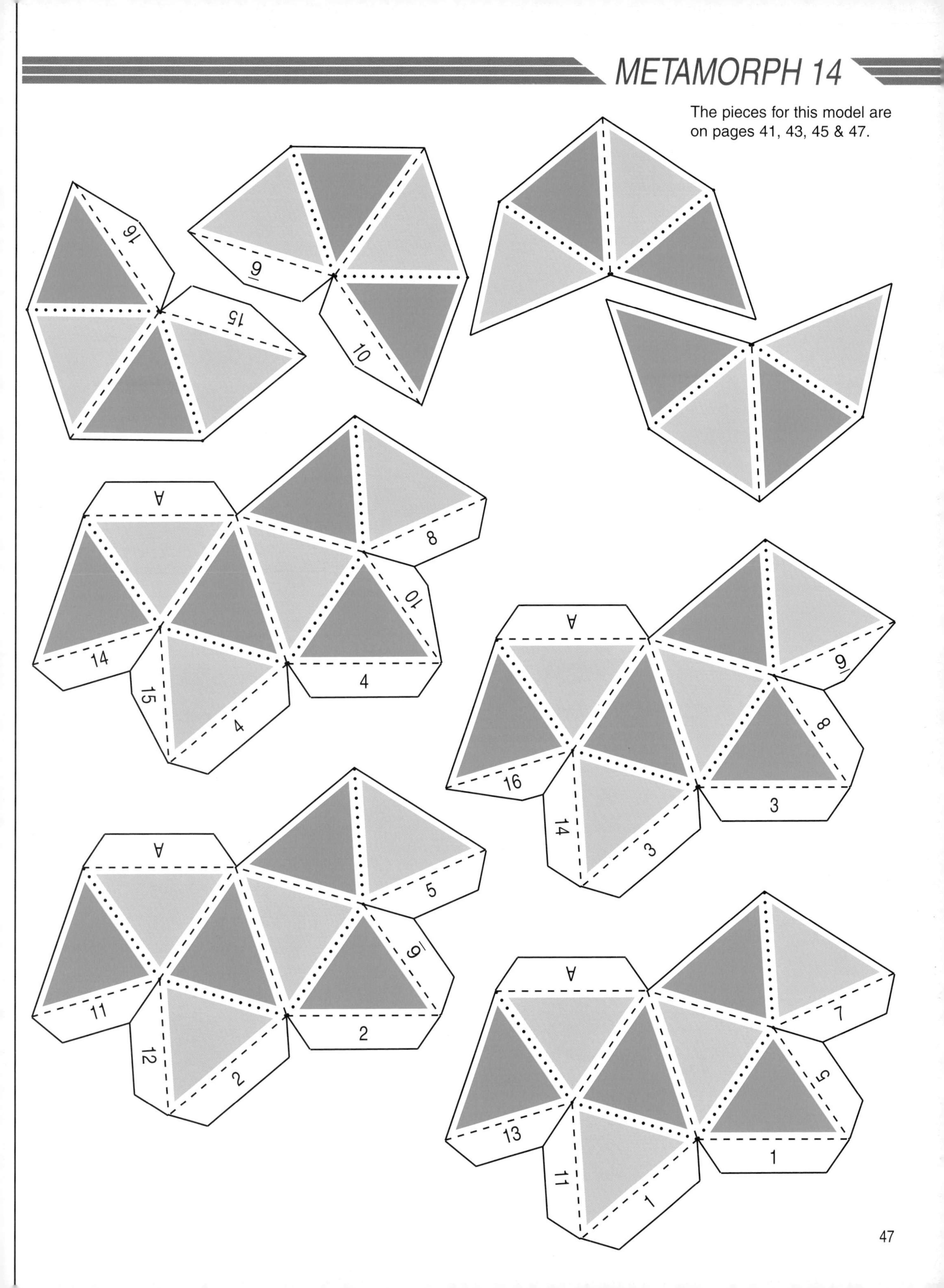

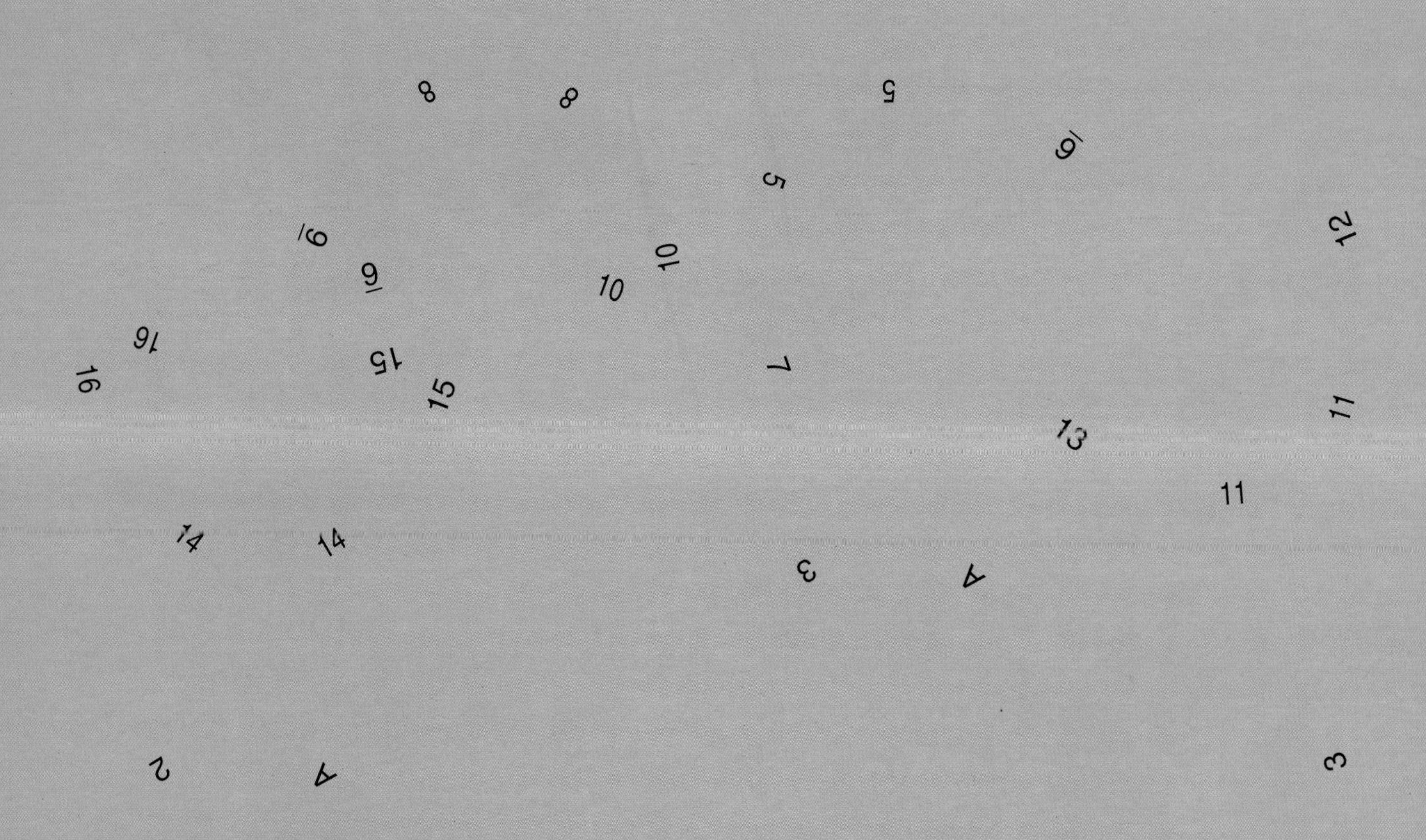